SpringerBriefs in Environment, Security, Development and Peace

ASEAN Studies

Volume 10

Series editor

Hans Günter Brauch

For further volumes:
http://www.springer.com/series/13033
http://afes-press-books.de/html/SpringerBriefs_ESDP_AS.htm

Nur Azha Putra · Eulalia Han
Editors

Governments' Responses to Climate Change: Selected Examples From Asia Pacific

 Springer

ENERGY
STUDIES
INSTITUTE

Editors
Nur Azha Putra
Eulalia Han
Energy Studies Institute
National University of Singapore
Singapore

ISSN 2193-3162   ISSN 2193-3170 (electronic)
ISBN 978-981-4451-11-6   ISBN 978-981-4451-12-3   (eBook)
DOI 10.1007/978-981-4451-12-3
Springer Singapore Heidelberg New York Dordrecht London

Library of Congress Control Number: 2013951766

Title Page Illustration: This map of Southeast Asia (2007) is in the public domain and was taken from the Perry-Castañeda Library Map Collection (Asia Maps) hosted by the University of Texas Libraries in Austin; Source is at: http://www.lib.utexas.edu/maps/middle_east_and_asia/southeast_asia_ref_2007.jpg.

Printed on acid-free paper

Springer is part of Springer Science+Business Media (www.springer.com)

Preface

This volume is a compilation of selected papers presented at a conference on *Policy Responses to Climate Change and Energy Security Post-Cancún: Implications for the Asia-Pacific Region's Energy Security*. The conference, which was held in March 2011 in Singapore, was organised by the Energy Studies Institute (ESI) at the National University of Singapore (NUS), Singapore.

The volume was conceptualised against the backgrounds of Asia's economic growth—which has sparked a fresh focus on the region's, and particularly China's, energy demand and consumption—and renewed vigour to address environmental issues amid the impasse in the international climate change regime. While developed countries have begun to embrace the importance of balancing the energy trilemma, namely energy security, economic competitiveness and environment sustainability, several developing countries in Asia are still reluctant to jeopardise their economic growths in the interest of mitigating climate change. Nevertheless, there are Asian states that are committed to reducing the impact of their growing economies on the environment. This volume therefore looks at the response of such governments with the intent of highlighting their strategies, so that it can serve as a model for other states.

The case studies selected in this volume represent some of the more influential countries in the region in terms of their rising prominence in world politics and the global economy. Understanding governments' responses to climate change in China, India, Indonesia and Singapore would provide an adequate overview of developments in Asia from the perspective of varying political, social and economic systems.

The introductory chapter by Nur Azha Putra and Eulalia Han, "Governments' Response to Climate Change: Issues, Challenges and Opportunities", discusses the conundrum governments in Asia face in delivering sustained economic growth while adopting policies that will help to mitigate climate change. The Chap. 1 seeks to highlight issues of contention in the climate change rhetoric, especially the disconnect that exists between government narratives and their policies. It also touches upon the opportunities presently available for multilateral cooperation in Asia, especially in addressing climate change.

In Chap. 2, "Engaging Emerging Countries: Implications of China's Major Shifts in Climate Policy", Gang He presents an overview of how China has integrated clean energy in its long-term goals of economic progress and engage emerging economies in global climate change efforts. His chapter articulates China's commitment to climate change and public health, and emphasises that international cooperation is fundamental to effectively address the non-traditional security threats facing states today. The Author also points out China's ambitions to be a world leader in clean energy and, perhaps more importantly, the implications for global climate change initiatives if China is able to progress towards a low-carbon economy while sustaining its economic growth.

In Chap. 3, "India's Efforts to Maintain and Enhance Energy Security While Reducing Greenhouse Gas Emissions", Harbans L. Bajaj highlights India's efforts in clean coal technology, nuclear energy and energy efficiency. The Author concludes that the main challenge for India, one of the world's fastest growing economies, lies in achieving its energy security and reducing its greenhouse gas emissions while improving the socio-economic condition of its citizens, especially for those who live below the national poverty line. As India is still largely dependent on coal to meet its rising energy demand, the country's commitment to address climate change through the successful implementation of clean technology and increasing the share of renewables in its energy mix will be crucial if it is to achieve its goals.

In Chap. 4, "Climate Change and Energy Security Post-Cancún: The Indonesia Perspective", Fitrian Ardiansyah, Neil Gunningham and Peter Drahos shed light on gaps between Indonesia's climate change and energy security objectives and its efforts on climate change. The Authors suggest that one of the most formidable challenges that Indonesia faces is the development of a consistent and sustainable domestic framework that can effectively guide the country's climate change efforts. For instance, political stability plays a significant role in the implementation of national policies and gaining the trust of its citizens would be crucial for the successful adoption of Indonesia's national strategies.

In Chap. 5, "Singapore's Policy Response to Climate Change: Towards A Sustainable Future", Authors Nur Azha Putra and Nicholas Koh offer insights on how the nation could socialise and therefore sustain its climate change policy that is largely founded on technology and market-based solutions by infusing its citizenry with the values of environmental citizenship. The Authors suggest that the socialisation of the country's climate change policies should begin at the grassroots level and driven by civil society and non-governmental organisations with the support of the government. Additionally, climate change should be taught in public schools and therefore introduced in the national education curriculum. To be effective, meaningful and sustainable, the Authors argue that ecological citizenship should be rooted in the nation's socio-political history, cultural and intellectual tradition.

In the concluding chapter on "Empowering the People: Towards the Inclusion of a Global Civil Society in a New Climate Change Regime", Eulalia Han supports the significance of developing a global civil society that would encourage cooperation between countries to address climate change in a concerted effort. She calls for the greening of nation-building and the formulation of adaptive policies, and stresses the potential part that a global civil society could play in narrowing the gaps between national and global initiatives towards climate change while prompting local communities from different countries to work towards a common goal.

Collectively, the chapters in this volume aim to discuss and introduce new ideas in the global climate change discourse while remaining sensitive to the economic and political issues and challenges in Asia-Pacific. We hope this volume will continue to spark fresh interests in the climate change discussions and dynamics of a region that is fast emerging as one of the most important actors in world politics.

Singapore, August 2013   Nur Azha Putra
                                                                                          Eulalia Han

Contents

Chapter 1
Introduction: Governments' Response to Climate Change: Issues, Challenges and Opportunities

Nur Azha Putra and Eulalia Han

Abstract As Asia continues to achieve new heights in political, social and economic development, this has inevitably led to a rise in energy demand and increased carbon dioxide emissions as well as a new sense of urgency to adopt sustainable practices towards this end. This chapter highlights the issues of contention in the climate change rhetoric, as articulated by governments, and the strategies that are in place to encourage sustainable development. It also discusses the opportunities and challenges for regional cooperation in Asia while contending that there is presently added prospect for engagement in the region, especially on issues such as addressing climate change, given the culture of pragmatism that is prevailing in the arenas of domestic and international relations.

Keywords Asia · Climate change · Regional cooperation

1.1 Introduction

Over the last few decades, Asia has seen strong and robust economic growth. However, with strong economic growth comes increased demand for energy, and this has in turn lead to higher greenhouse gas (GHG) emissions. According to the World Energy Outlook 2012,[1] Asian countries will remain undeterred by rising

---

[1] WEO, "World Energy Outlook 2012", at: http://www.iea.org/publications/freepublications/publication/English.pdf (31 July 2013).

N. A. Putra (✉) · E. Han
Energy Studies Institute (ESI), National University of Singapore (NUS),
29 Heng Mui Keng Terrace, Block A, #10-01, Singapore 119620, Singapore
e-mail: azha@nus.edu.sg
URL: http://www.esi.nus.edu.sg/about-us/our-researchers/nur-azha-putra

E. Han
e-mail: esihne@nus.edu.sg
URL: http://www.esi.nus.edu.sg/about-us/our-researchers/dr-eulalia-han

N. A. Putra and E. Han (eds.), *Governments' Responses to Climate Change:*
Selected Examples From Asia Pacific 10, SpringerBriefs in Environment, Security,
Development and Peace, DOI: 10.1007/978-981-4451-12-3_1, © The Author(s) 2014

prices and continue to rely on fossil fuels such as oil, natural gas, and coal to power their economies, at least in the foreseeable future. As it stands, Asian economies simply do not possess cheap and viable fossil fuel alternatives that can be deployed without risking their economic development and energy security. Nevertheless, Asian economies generally recognise that their continued reliance on fossil fuels hinder sustainable development.

Climate change is occurring and its effects have begun to take its toll, especially in the Asia Pacific region. Rising sea levels have had devastating effects on several Asia–Pacific islands. Elsewhere in the Asia Pacific, governments are struggling to mitigate frequent occurrences of flash floods due to frequent short bursts of heavy rainfalls and the loss of landmass due to diminishing coastlines. The world is getting warmer, too. The Arctic ice is melting, and there are fears that this could further lead to increased sea water levels. Rising sea water levels can not only diminish coastlines but also intrude into fresh water streams and reservoirs, thereby rendering fresh water sources inconsumable. Mountain glaciers are not forming quickly enough, and this affects the water security of many nations, particularly those in the South Asia region where fresh water supply is dependent on the glaciers in the Himalayas. Prolonged dry spells have affected agriculture. The food security of many communities is threatened by barren lands and the shortages or absence of rainfall. Other than its impacts on the environment, climate change poses security threats to human collectivities, too. Rivers that used to be a lifeline for coastal communities are drying up, forcing people to migrate inland and intrude into the space of other villages. In instances of forced migration, there have been reported cases of conflicts between communities. Population displacement and forced migrations are just two examples of the impact of climate change on human collectivities. In short, the threat of climate change has, in recent years, evolved to one that presents clear and present danger to the states.

With growing recognition of the threats of climate change, the international community has upped the ante. Recent rounds of the United Nations Framework Convention on Climate Change (UNFCCC) negotiations have seen a large improvement in the participating states' willingness to implement policies that aim to mitigate climate change. Reducing GHG to the extent that it does not negate energy security and economic growth appears to be a common feature in today's political lexicon. The international community has displayed a renewed vigour towards diversifying its energy mix to include environmentally friendly fuels such as hydro, nuclear, solar, biofuels and thermal options. Vigour, however, does not necessarily imply commitment to progress. Progress is still slow. Many economies continue to rely on oil and gas, and any attempt to mitigate climate change has to be considered within the larger framework of economic development. States, in general, are reluctant to implement expensive green technologies, reduce car populations, impose hefty green taxes and diversify their national energy mix to include a larger percentage share of expensive non-fossil fuels in the interest of protecting the economy.

As a growing middle class in consumerist-driven economies such as China, India, Vietnam, Thailand, Malaysia, Singapore and Indonesia have driven the demand for more goods and services, this has in turn driven the need for more energy. Although Asia is rich in natural resources, countries in the region still predominantly rely on fossil fuels rather than clean energies such as hydro, solar, nuclear and biofuel sources. Consumers' choices for goods and services are still driven by personal preferences that are not rooted in the interests of climate change or sustainable development. This is unlikely to change drastically in the absence of green and environmentally friendly policies in the region such as carbon and green taxes. In the interest of sustainable development and mitigating climate change, governments therefore will have to intervene and influence consumer behaviours. This implies that governments too must implement policies at the national level that reflect the global threat of climate change. However, national policies for countries that are part of the global economy are unlikely to work in the absence of international cooperation. Addressing climate change requires a concerted effort, as neither are its effects constrained by geography nor does it only affect countries with extensive mining and production capabilities. Multilateral cooperation is therefore necessary, but can only be effective when governments' rhetoric and strategies are consistent and in line with the long-term goals of the international community. However, this is far easier said than done because not all countries are at the same level of political, social and economic development.

As a whole, Asia has experienced dynamic and sustained economic progress beginning in the last century, particularly in post-colonial Southeast Asia. Asian countries and China, in particular, are beginning to assert their economic presence in the quest for energy security around the world. Kishore Mahbubani, Dean and Professor in the Practice of Public Policy, Lee Kuan Yew School of Public Policy, National University of Singapore, Singapore, contends that the international community will see an end to Western domination of world history and the rise of Asian entrepreneurship, be it in business, innovation or policy-making (Mahbubani 2008: 9). What does this mean for Asia, which comprises countries with competing histories and ideologies and, more specifically, what does this mean for regional cooperation when it comes to addressing global issues such as climate change amid impressive political, social and economic development?

1.2 Climate Change: Then and Now

Discussions surrounding international environmental issues first emerged in the nineteenth century in the context of managing resources and in the face of recognition that these issues were inherently global (Greene/Owen 2005: 453). The first international treaty on the protection of flora was signed in 1889, which was followed by the 1902 Convention for the Protection of Birds Useful to Agriculture (Greene/Owen 2005: 453). In 1945, the preservation of natural resources was included as part of the mandate of the Food and Agriculture Organization (FAO) of the United Nations (UN)[2] and, in 1946, an International Whaling Conference was established to monitor the rate of whaling activities.[3] It was only in the 1960s, however, that attention was accorded to the problem of pollution and the importance of preserving the environment through the publication of *Silent Spring* by Rachel Carson (1962).

In 1972, the UN Conference on the Human Environment in Stockholm was organised to emphasise "a common outlook and for common principles to inspire and guide the peoples of the world in the preservation and enhancement of the human environment".[4] During the 1970s and 1980s, in conferences held from Stockholm to Rio, environmental issues became institutionalised in formal treaties (Greene/Owen 2005: 454). More significantly, this gave way to increased participation of non-governmental organisations and raised concerns over the relationship between economic progress and development, mainly being experienced by developed countries (Greene/Owen 2005: 456–459). In 1987, the Brundtland Commission saw the emergence of discussions surrounding the importance of achieving sustainable development (WCED 1987) and, since then, the debates surrounding climate change and environmental issues have transformed from one of recognition of the problem to that of mitigation and, finally, adaptation.

As Asia develops, its governments are acutely aware that its progress is having a profound impact on the environment. For example, China is the largest producer and consumer of coal, the most pollutant fossil fuel (BP 2012: 32–33). However, as Chief Economist of the International Energy Agency (IEA) Fatih Birol noted, China also has the most environmentally conscious government (Mahbubani 2008: 191).

Going forward, 2015 could see the UNFCCC reach a new comprehensive agreement on addressing environmental issues. The Kyoto Protocol,[5] which was first enshrined in 1997, made a distinction between developed and developing countries, where the former were required to cut emissions but the latter were not. By 2020, the Kyoto Protocol "will be replaced by a single legal agreement that ends the outdated binary distinction between 'developed' and 'developing' countries and requires all to make commitments commensurate with their level of development".[6] Apart from international climate change efforts, one of the most significant changes to come from the Doha Climate Change Conference in November 2012 was the recognition of the need "to address loss and damage associated with

---

[2] FAO, "A Short History of FAO", at: http://www.fao.org/about/en/ (10 July 2013).

[3] IWC, "Welcome to the IWC", at: http://iwc.int/home (10 July 2013).

[4] UNEP, "Declaration of the United Nations Conference on the Human Environment", at: http://www.unep.org/Documents.Multilingual/Default.asp?documentID=97&ArticleID=1503 (10 July 2013).

[5] UNFCCC, "Kyoto Protocol", at: http://unfccc.int/kyoto_protocol/items/2830.php (15 July 2013).

[6] Jacobs, Michael, "The Doha Climate Talks were a Start, But 2015 will be the Moment of Truth", in: *The Guardian* (10 December 2012), at: http://www.guardian.co.uk/commentisfree/2012/dec/10/doha-climate-talks-global-warming (28 June 2013).

climate change impacts in developing countries that are particularly vulnerable to the adverse effects of climate change to enhance adaptive capacity".[7]

One of the main dilemmas of international conventions, however, is to balance the needs of developing countries, which are only just beginning to reap the benefits of industrialisation and the free market, and that of developed countries that were part of the Industrial Revolution and whose industries have dominated the last 200 years. Mahbubani (2008: 188) notes that "the greenhouse effect that we are worrying about is not being *caused* by current emissions, even though it is being aggravated by them. The fundamental cause is the stock of emissions we have accumulated, especially in the last two centuries since the Industrial Revolution". While the developed countries are now concerned about the consequences of their economic growth, with some expressing willingness to reduce their emissions, imposing the same curbs on developing countries might pose a huge obstacle to the latter's social development and economic progress.

Nonetheless, there is a global consensus that climate change represents a real threat. Addressing this will require a concerted effort between states and their respective bodies politic. In this context, it is necessary that any global initiative that seeks to address climate change balances the delicate relationship between the needs of developed and developing countries, and engages civil society and non-state actors in its policy formulation. Understandably, the liberties found in democracies that encourage the participation of civil society in such initiatives might not be shared in all countries. However, the signs are there to suggest that even the most closed societies are gradually opening up—not simply as a result of free market economics but because the challenges of the day, especially climate change, require the cooperation and involvement of all those that are affected.

1.3 The Significant Role of Governments in Addressing Climate Change

By their very nature, environmental issues transcend state boundaries. Questions of overexploitation have also been brought to the fore as "the processes leading to over-exploitation and environmental degradation are intimately linked to broader political and socioeconomic processes, which themselves are part of a global political economy" (Greene/Owen 2005: 453). Therefore, global environmental issues not only affect the physical living conditions of the international community but also constrain the social and economic progress of communities that have been disadvantaged by the process of overexploitation.

It is important to note that "international environmental problems are rarely caused by deliberate acts of national policy, but are rather unintended side-effects

---

[7] UNFCCC, "Doha Climate Change Conference—November 2012", at: http://unfccc.int/meetings/doha_nov_2012/meeting/6815/php/view/decisions.php (28 June 2013).

of broader socio-economic processes" (Greene/Owen 2005: 457). States are not the only actors in this process; financial institutions, private companies and individuals also play crucial roles. Acknowledging the relationship between production and consumption at any level—be it at the state, institutional or individual level—is important, and attributing sole responsibility to any one level will only serve to enhance the problem. However, because climate change is a global problem and sovereign states are the only recognised representations of citizens under international law, governments should naturally shoulder a larger proportion of efforts targeted at addressing environmental issues.

The way citizens think about and respond to issues is reflective of the environment that has nurtured those instincts. This environment is a combination of their participation in social networks, their understanding of their country's history and, most importantly, their experiences in formalised education institutions. The state plays a significant role in shaping the characters of these institutions and networks, as determined by its role in the development of the national curriculum, the allocation of national resources, and the priorities and commitments articulated by the country's leaders. There may be instances, however, whereby the state's rhetoric and strategies highlight a disconnect in terms of what the government purports it is committed to doing and its actual level of commitment. One should bear in mind that this does not necessarily mean that citizens should be sceptical of such commitments, as it is genuinely a difficult task for governments to apply the same clarity and consistency to all policies. Then again, for issues such as addressing climate change where a global concerted effort is needed, governments should develop and adjust domestic institutions and policies so that they are able to respond quickly to complex issues in an increasingly complex world.

1.4 Climate Change and Regional Cooperation in Asia: Opportunities and Challenges

Plaudits surrounding the steadfast rise of Asia and its firm belief in the importance of pragmatism when practising domestic and international relations are frequently accompanied by critiques that question Asia's rule of law, pseudo-democratic practices and violations of human rights (Bauer et al. 1999). Where the potential for the Asian region to cooperate on addressing the various challenges surrounding energy and environmental issues is concerned, it has also been suggested that the major obstacles that Asia faces are four-pronged: (a) Asia's energy requirements are varied, and this makes arriving at a consensus on climate change strategies difficult; (b) it has been suggested that "Asian governments have a traditionally narrow conception of national security and the persistent fear of relative gains has stymied the scope of cooperation over key energy issues" (Len et al. 2009b: 9); (c) the political culture of the Asian region usually places emphasis on form over substance, thus slowing down the processes of implementation and institutionalisation of policies and practices; and, (d) the governments of Asian countries lack political trust in one another (Len et al. 2009a: 15).

For instance, it has been noted that China and Japan see "energy security in security terms, and not economic terms", "within narrow security calculations and in the framework of strategic thinking", and therefore "regard each other as competitors, instead of partners with potential for cooperation" (Choo/Jaewoo 2009: 57). The often-tense relationship between the two countries might have also been influenced by the events of World War II and China overtaking Japan as Asia's largest economy. In Southeast Asia, others contend that while the completion of the Trans-ASEAN (Association of Southeast Asian Nations) Gas Pipeline (TAGP) and the ASEAN Power Grid (APG) could improve ASEAN's overall energy security, various challenges including technical difficulties, financial constraints, and the differences in the regulatory and legal frameworks of the various ASEAN countries could hinder their full development.[8] In addition, given South Asia's uneasy past (especially the choppy relations between India and Pakistan) and the current competition between rising powers in the region, regional cooperation on the energy front has been weak and vulnerable (Raja Mohan 2009: 83).

Amitav Acharya, Professor of International Relations at the School of International Service, American University, contends that the Asia–Pacific region has adopted the main principles of the 'ASEAN Way' (Acharya 1997: 319). Therefore, it is appropriate to infer that, according to Acharya, Asia "is witnessing a cautious, pragmatic, informal, gradualist and consensus-seeking approach to multilateral institution-building", and multilateralism "remains constrained by the primacy of state interests and conflicting conceptions of regional identity" (Acharya 1997: 342). However, he notes that 'pragmatism', 'consensus', 'open regionalism' and 'soft regionalism', as practiced in Asia, could be alternatively interpreted in another manner. The significance of these approaches, he notes, "may lie in their value as ways of reconciling conflicting state preferences and finding a common ground out of differing economic, political and strategic priorities among members" (Acharya 1997: 343).

Mahbubani contends that, most importantly, the approach practised in Asia has worked well for countries in the region.[9] He compares the European Union (EU) and ASEAN, and notes that the EU is always trying to be ideologically consistent and hence it will not admit any member states unless they have met a whole series of standards. It wants to have a club of "people like us". The fundamental geopolitical mistake that the EU has made is its failure to understand that the vast majority of the world's population does not consist of 'people like us'. According to Mahbubani, to only admit or deal with 'people like us' creates a club of

---

[8] Nugroho, Hanan, "ASEAN Energy Cooperation: Facts and Challenges", in: *The Jakarta Post* (19 May 2011), at: http://www.thejakartapost.com/news/2011/05/19/asean-energy-cooperation-facts-and-challenges.html (10 July 2013).

[9] Mahbubani, Kishore, "Can the EU Learn Lessons from ASEAN?", in: *Europe's World* (2012), at: http://www.mahbubani.net/articles%20by%20dean/can-the-eu-learn-lessons-from-asean.pdf (10 July 2013).

exclusion. By contrast, ASEAN has set the gold standard globally by becoming the club of 'inclusion' and not 'exclusion'. Indeed, for decades, ASEAN was subjected to derision by Western thinkers because it had admitted a flawed military regime in Myanmar as a member to the ASEAN.[10]

Asia's regional cooperation should not only focus on its economic achievements but also reflect on the other aspects of domestic and international affairs that do not necessarily require huge reserves to address. These areas of significance include encouraging a lively civil society, openness towards engaging countries that Asian countries have had tumultuous relationships with in the past and a serious commitment to addressing environmental issues that transcend state boundaries. While there remain obstacles to regional cooperation, the culture of pragmatism within Asia has the potential to allow the region to effectively cooperate on most issues, especially addressing climate change, the effects of which all countries are experiencing.

1.5 Conclusion

Moving forward, any governmental policies to mitigate climate change may remain ineffective if the policy of inclusion is limited to the level of state-to-state cooperation. Non-state actors should be included at all policy levels, either as consultative partners or practitioners, as climate change mitigation and adaptation strategies would be implemented more effectively through state and society partnership. Such a partnership may be the answer to questions regarding issues of policy and implementation gaps, especially in situations where well-designed and crafted policies fail to produce the results they are intended to achieve in the first place.

References

Acharya, Amitav, 1997: "Ideas, Identity, and Institution-Building: From the 'ASEAN Way' to the 'Asia-Pacific Way'?", in: *The Pacific Review*, 10,3: 319–346.
Bauer, Joanne R.; Bell, Daniel A. (Eds.), 1999: *The East Asian Challenge for Human Rights* (Cambridge: Cambridge University Press).
BP, 2012: *BP Statistical Review of World Energy June 2012* (London: BP Statistical Review of World Energy).
Carson, Rachel, 1962: *Silent Spring* (Harmondsworth: Penguin).
Choo, Jaewoo, 2009: "Northeast Asia Energy Cooperation and the Role of China and Japan", in: Len, Christopher; Chew, Alvin (Eds.): *Energy and Security Cooperation in Asia: Challenges and Prospects* (Sweden: Institute for Security and Development Policy): 41–59.

---

[10] Ibid.

Greene, Owen, 2005: "Environmental Issues", in: Bayliss, John; Smith, Steve (Eds.): *The Globalization of World Politics: An Introduction to International Relations*, 3rd ed. (Oxford: Oxford University Press): 452–78.

Len, Christopher; Chew, Alvin, 2009a: "Executive Summary", in: Len, Christopher; Chew, Alvin (Eds.): *Energy and Security Cooperation in Asia: Challenges and Prospects* (Sweden: Institute for Security and Development Policy): 15–18.

Len, Christopher; Chew, Alvin, 2009b: "Preface", in: Len, Christopher; Chew, Alvin (Eds.): *Energy and Security Cooperation in Asia: Challenges and Prospects* (Sweden: Institute for Security and Development Policy): 9–13.

Mahbubani, Kishore, 2008: *The New Asian Hemisphere: The Irresistible Shift of Global Power to the East* (New York: PublicAffairs).

Raja Mohan, C., 2009, "The Role of Energy in South Asian Security", in: Len, Christopher; Chew, Alvin (Eds.): *Energy and Security Cooperation in Asia: Challenges and Prospects* (Sweden: Institute for Security and Development Policy): 83–102.

WCED, 1987: *Our Common Future* (Oxford: Oxford University Press).

Abbreviations

APG	ASEAN Power Grid
ASEAN	Association of Southeast Asian Nations
BAU	'Business as usual'
EU	European Union
FAO	Food and Agriculture Organization
GHG	Greenhouse gas
IEA	International Energy Agency
IWC	International Whaling Commission
TAGP	Trans-ASEAN Gas Pipeline
UN	United Nations
UNEP	United Nations Environment Programme
UNFCCC	United Nations Framework Convention on Climate Change
WCED	World Commission on Environment and Development

Chapter 2
Engaging Emerging Countries: Implications of China's Major Shifts in Climate Policy

Gang He

Abstract Engaging developing countries, especially emerging counties such as Brazil, Russia, India and China, is a central challenge for international climate policy. While the world is still debating the outcome of international climate negotiations, China has been quietly developing its leadership in clean energy and moving towards a low-carbon economy and society. By end 2009, for instance, China had become the world's leading new wind installer, solar exporter and new nuclear capacity constructor, in total, as well as the world's top investor in clean energy, while at the same time becoming the world's biggest carbon dioxide emitter. China's move towards a low-carbon economy will have big implications on global energy and climate policy. China's transition typifies a unique opportunity for developed economies to engage emerging economies in the global climate efforts.

Keywords Aligning interests · China · Climate policy · Engaging · Incentives · Shift

2.1 Introduction

Engaging developing countries, especially emerging economies such as Brazil, Russia, India and China, is central to any effort that addresses global climate change. The current climate negotiations are affected by a fundamental conflict of

G. He (✉)
Energy and Resources Group (ERG), University of California, Berkeley, 310 Barrows Hall, Berkeley, CA 94720, USA
e-mail: ganghe@berkeley.edu
URL: http://www.ganghe.net

N. A. Putra and E. Han (eds.), *Governments' Responses to Climate Change:
Selected Examples From Asia Pacific* 10, SpringerBriefs in Environment, Security,
Development and Peace, DOI: 10.1007/978-981-4451-12-3_2, © The Author(s) 2014

interest between the developed and developing worlds.[1] The developed countries
contend that unless the developing world and emerging economies,[2] in particular,
participate in an effective way, there cannot be a concerted effort in tackling
climate change, as these countries significantly contribute to global emissions
(IEA 2010: 85).[3] This stand was reflected in the climate talks that peaked at
Copenhagen. Indeed, many experts believe that the real challenge is how to
effectively engage the participation of emerging countries in emission limits before
a global legally binding agreement is possibly reached (Giddens 2009: 849;
Keohane/Victor 2011: 420; Victor 2011: 210).

Even as the world continues to debate the outcome of international climate
negotiations and with the developed countries remaining reluctant to take deep
cuts, China has been quietly developing its leadership in clean energy and moving
towards a low-carbon economy and society. By end 2009, China had become the
world's number one new wind capacity installer, solar exporter and new nuclear
capacity constructer. China was, in effect, not only the world's biggest carbon
dioxide (CO_2) emitter but also its top investor in clean energy.

Major strategy shifts have been evident in China's climate policy in recent
years. The leadership sees mitigating climate change as consistent with China's
other priorities such as energy security and pollution control. What is more, the
government views this as a new economic opportunity where China could compete
with technologically advanced countries, especially those that have dominated
conventional sources of energy for years. This major shift has motivated a series of
political, economic and institutional reforms in the country.

Should China, the world's top energy consumer and CO_2 emitter, move towards
a low-carbon economy, there would be important implications for global energy
and climate change policies. How would China manage its transformation to a
lower-carbon economy? And, what role would China play in developing renewable and clean technology solutions for the rest of the world? These are without
doubt questions that will have profound impact on the world energy and climate
landscape for years to come.

This chapter discusses China's changing role in the global climate change
negotiations, from Copenhagen to Durban, and then explore the central drivers of
its major policy shifts, highlighting also its domestic constrains at the same time.

---

[1] This deadlock in international climate negotiations between developing and developed
countries has been showcased by various experts, including Figueres (2010: 3), Victor (2011),
Hale et al. (2013: 251), Gao and Wu. See also Gao, Feng, "The Road Ahead from Bali", in:
China Dialogue (21 January 2008), at: https://www.chinadialogue.net/article/show/single/en/
1640-The-road-ahead-from-Bali (17 July 2013): 2; Wu, Changhua, "Breaking the Climate
Deadlock" (in Chinese) (1 June 2008), at: http://classic.theclimategroup.org.cn/projects/
breaking_the_climate_deadlock/ (7 March 2011).

[2] Brazil, Russia, India, China, and South Africa (BRICS) are normally considered merging
economies.

[3] IEA research shows that, by 2030, 80 percent of carbon emission growth will be from the
developing world.

This chapter concludes by examining the broader implications of China's policies for global energy and climate change strategies.

2.2 Major Shifts in China's Climate Policy

Issued in June 2007 by the National Development and Reform Commission (NDRC), China's National Climate Change Programme[4] was the first comprehensive policy initiative that explained China's climate change policy. Over the years, the Chinese government has developed several policy papers on the issue,[5,6,7] and there have been fundamental shifts in China's climate change policy.

First, the shift in its leadership's thinking and public awareness of the problem mean that China's climate change policy now converges at a point where such change is actually made possible. At the highest level, for instance, China's President Hu Jintao in his remarks to the 64th United Nations General Assembly stressed that "Climate change, food security, energy and resource security and public health security are all global challenges and no country is immune from them",[8] reaffirming China's commitments on climate change. Climate change has become a much-discussed topic in major international, bilateral and multilateral meetings as also conferences involving Chinese leaders. In recent years, climate diplomacy has played an increasingly key role in China's non-conventional diplomacy (Wang 2011: 68).

Similarly, at the ground level too, civil society organisations, especially non-governmental organisations, and youth groups are playing a key role in promoting public awareness of climate change in China (He 2007: 77). These non-state actors constitute China's socioeconomic climate change strategy and are a central component of China's diplomatic relations with the international community.

Next, the government's official position during negotiations has been more open, flexible and proactive lately. Described by Professor Haibin Zhang of Peking

---

[4] NDRC, "China's National Climate Change Programme", June 2007, at: http://www.ccchina.gov.cn/WebSite/CCChina/UpFile/File188.pdf (7 March 2011).

[5] NDRC, "China's Policies and Actions for Addressing Climate Change—The Progress Report 2009", November 2009, at: http://www.ccchina.gov.cn/WebSite/CCChina/UpFile/File571.pdf (8 April 2011).

[6] NDRC, "China's Policies and Actions for Addressing Climate Change—The Progress Report 2010" (in Chinese), November 2010, at: http://www.ccchina.gov.cn/WebSite/CCChina/UpFile/File927.pdf (8 April 2011).

[7] NDRC, "China's Policies and Actions for Addressing Climate Change (2012)", at: http://www.ccchina.gov.cn/WebSite/CCChina/UpFile/File1324.pdf (1 December 2012).

[8] Hu, Jintao, "Statement by President Hu Jintao at the General Debate of the 64th Session of the United Nations General Assembly", at: http://www.china-un.org/eng/hyyfy/t606150.htm (28 April 2011).

University as "what has remained unchanged is that China still refuses to take a binding GHGs emissions reduction commitment, while China's attitude towards the international climate negotiations has become more flexible and cooperative", this stance has "remained unchanged" till today (Zhang 2006: 276).

Finally, an institution for addressing climate challenge has been established and strengthened. The establishment of the Leading Group on Energy Efficiency and Emission Reduction and Confronting Climate Change in 2007[9] meant that climate change was now a key agenda for the State Council. In 2009, the Division of Climate Change[10] was established under the NDRC, China's central economic management and regulation body.

2.3 Climate Change as a New Economic Opportunity: Understanding China's Participation in Global Climate Change Efforts

It is important that we understand China's various initiatives as part of the global climate change efforts.

2.3.1 What Is China Doing?

First and foremost, energy efficiency remains at the top of China's energy and climate change policy agendas (Price et al. 2011: 2165). In 2006, China announced a national programme, called China's 'Energy Conservation and Emission Reduction' programme, that aimed to achieve a 20 percent energy intensity improvement based on 2005 levels in its 11th Five-Year Plan.[11] By 2010, China recorded a 19.1 percent reduction in its energy intensity, meaning that the total energy savings during the 11th FYP was 600 million tons of coal equivalent (tce) while emission reduction was at 1.5 billion tons of CO_2.[12] From 2006 to 2009,

---

[9] State Council, "Notice on the Establishment of Leading Group on Energy Efficiency and Emission Reduction and Confronting Climate Change" (in Chinese), No. 18, 2007, at: http://www.gov.cn/zwgk/2007-06/18/content_652460.htm (28 April 2011).

[10] For more information, see Division of Climate Change, "Organization structure" (in Chinese), at: http://qhs.ndrc.gov.cn/jgsz/default.htm (28 April 2011).

[11] At the time when the policy was introduced, emission meant sulphur dioxide (SO_2) and chemical oxygen demand (COD); carbon dioxide (CO_2) was not yet included in the term. See NDRC, "National Economy and Social Development 11th Five-year Plan" (in Chinese), at: http://www.gov.cn/ztzl/gmjj/ (28 April 2011).

[12] NDRC, "Energy Efficiency and Emission Reduction Progress Report" (in Chinese), at: http://www.ndrc.gov.cn/xwfb/t20110310_399044.htm (28 April 2011).

China shut down 60 GW of small and inefficient power plants,[13] accounting for more than three times the total capacity of the Three Gorges Hydro power plant. Then again in its 12th FYP, China proposed a new target of 16 percent energy intensity reduction and 17 percent emission reduction based on 2010 levels.[14]

A second area of focus for China has been clean energy. China is undergoing a clean energy revolution and is the world's leader in renewable energy and clean energy development and deployment.[15] China tops global green stimulus investment, with total investments amounting to US$218 billion by end 2010 following the 2008 global economic recession. It is predicted that China's wind power consumption capacity by 2015 will be 90 GW and, by 2020, this figure will stand at 150 GW.[16] High-speed rail is another example. By 2020, China is to build a national high-speed rail system spanning more than 16,000 km.[17] It has been projected in the National Climate Change Plan (2011–2020) that investment to the tune of 10 trillion RBM (or nearly US$1.5 trillion) will be needed in order to achieve China's 40–45 percent emission intensity target.[18]

The above examples highlight some of China's ambitious targets for clean energy development. There is certainly an opportunity for China to compete in the global clean energy market. Moving towards a lower-carbon society will not only allow China to be more economically competitive but also lay the ground for international cooperation on clean energy, as China's scale and speed of development with respect to such technology has been significant.[19]

---

[13] NDRC, "Power Industry Annual Statistical Report 2009" (in Chinese), at: http://www.ndrc.gov.cn/jjxsfx/t20100713_360615.htm (28 April 2011).

[14] NDRC, "National Economic and Social Development 12th Five-Year Plan" (in Chinese), at: http://www.ndrc.gov.cn/fzgh/ghwb/gjjh/P020110919592208575015.pdf (28 April 2011).

[15] The Climate Group, "China's Clean Energy Revolution" (10 April 2008), at: http://www.theclimategroup.org/_assets/files/Chinas_Clean_Revolution.pdf (28 April 2011); Gordon, Kate; Wong, Julian L., "Out of the Running? How Germany, Spain, and China are Seizing the Energy Opportunity and Why the United States Risks Getting Left Behind" (4 March 2010), at: http://www.americanprogress.org/issues/green/report/2010/03/04/7386/out-of-the-running/ (16 June 2013).

[16] State Grid, "State Grid White Paper on Promoting Wind Energy Development", at: http://www.sgcc.com.cn/ztzl/Newenergy/gsbd/09/281068.shtml (27 September 2012).

[17] The plan was adopted in 2004, with an original target at 10,000 km, which was revised in 2011 by the ministry. See Ministry of Railway, "Middle and Long-term Plan of National Railway Networks", at: http://www.gov.cn/ztzl/2005-09/16/content_64413.htm (28 April 2011).

[18] Li, Yanzheng. "Low Carbon Investment May Reach 10 Trillion" (in Chinese), in: *Shanghai Security News* (28 April 2011), at: http://news.hexun.com/2011-04-28/129110260.html (29 April 2011).

[19] Kammen, Daniel; He, Gang, "Critical Test for Sino-US Ties: Clean Energy", in: *China Daily* (21 January 2011), at: http://www.chinadaily.com.cn/usa/china/2011-01/21/content_11895619.htm (21 March 2011).

2.3.2 What Is China Not Doing?

First, unlike in the United States, where climate change sceptics have been contributing towards a 'climate of doubt', China has since 2008 released two national assessment reports on climate change that have driven consensus among the scientific community and policy groups (MOST et al. 2007, 2011). Complimenting the government's various initiatives, the Party's central leaders, or the Politburo, have been inviting top climate change scientists and scholars to speak on China's strategies to tackle climate change.[20]

Second, China is playing a proactive role in tackling climate change. Prior to Copenhagen, China had always blamed the United States for the impasse in global climate negotiations, commonly referred to as the US-China climate deadlock (Lieberthal/Sandalow 2009: 67; Victor 2011: 61). But, since Copenhagen, China has seized the initiative,[21] marking a significant step forward in the international climate regime.

Third, China is not going to accept any physical cap on its emissions in the short term before peak emissions. The targets that China offers are mostly intensity based, an approach that makes sense for an emerging economy such as itself. Even in the most optimistic scenario, China will not be able to peak its energy consumption and so too its carbon emission by around 2025–2030 (Fridley et al. 2011: 30; Jiang et al. 2010: 214; Wang/Watson 2010: 3537; Zhou et al. 2011: 37).

2.4 China's Core Interests in Energy Supply and Its Climate Policy Implications

The main priority for the Chinese government is to sustain its economic growth and maintain its macrosocioeconomic stability. An important query that arises here is—what has this to do with energy and climate? 'Energy security',[22] as far as the Chinese government is concerned, is to ensure not only that the security of supply will not impede its economic growth but also that the price of energy remains sufficiently low in the interests of maintaining social stability. Should this not be achieved, the fluctuation of energy prices would have a detrimental impact on its national economy and social stability.

China is highly dependent on coal, as 70 percent of its primary energy and about 80 percent of its electricity is generated from this source (NBS 2010: 269). However, without carbon capture and sequestration (CCS), coal would be a

---

[20] "Addressing Climate Change Properly: The 19th Group Study of the Politburo" (24 February 2010) (in Chinese), at: http://politics.people.com.cn/GB/1024/11012529.html (28 April 2011).

[21] NDRC, "China's Official Position Statement on Climate Change" (in Chinese), at: http://www.sdpc.gov.cn/zcfb/zcfbqt/2009qt/t20090521_280387.htm (28 April 2011).

[22] Zhang, Guobao, "Current Energy Situation: Opportunity in Crisis" (in Chinese) (30 December 2008), at: http://energy.people.com.cn/GB/71893/8599677.html (8 April 2011).

substantial source of pollution. For China, a more realistic issue, where coal is concerned, is the safety of its coal mining operations and its need to ensure a sustainable supply that could otherwise disrupt the country's transportation industry and coal power generation sector (Morse et al. 2009: 2; Peng 2009: 10; Rui et al. 2010: 15). China plans to reduce its dependency on coal to below 40–50 percent by 2050 (CAE 2011: 10), thus underscoring the importance of the diversification of its energy supplies and an improvement of efficiency to its energy security. China also recognises the potential for developing leadership in energy technology manufacturing and exports. Reflecting this change, 'innovation in China' has become a more attractive tag than 'made in China'.

Looking ahead, China's climate change and environmental policies need to address its pollution,[23] which should have been a top priority on China's environmental agenda for some time. For example, a series of disastrous incidents seen in recent years such as air pollution, acid rain, water pollution and heavy metal poisoning are believed to be a result of accumulated environmental pollution and degradation over 30 years of rapid economic development (FON 2011: 3). As coal is one of the main sources of air pollution in China, addressing the domestic air pollution scenario will require more effective cooperation between central and local authorities.

Although climate security was not always at the top of the country's policy agenda, it has, in recent years, become increasingly prominent in the country's policy matrix. The government has realised that China will not be excluded from the threats of climate change. According to China's National Assessment of Climate Change, the country's agriculture industry will be affected by extreme climate by as much as 10 billion RMB each year (MOST 2007). This will imply that, together with the challenges of water security and food security, the Chinese government will need to address climate security in the context of the energy-water-food-climate nexus.

From the country's developmental perspective, climate change has already become embedded in the government's sustainable development strategy. However, different policies and technologies need to be prioritised in order to achieve an optimised allocation of resources. Chinese priorities, then, can be broadly divided into four groups, as shown in Fig. 2.1. Better understanding of where the interests and priorities of China lie relative to those of the world, as a whole, will help us to better align these for the common good.

2.5 Constrains and Uncertainties

As we discuss China's initiatives on the climate change front, it is important to highlight here some of the domestic constrains that could hinder these efforts.

---

[23] MEP, "2009 Report on the State of the Environment in China", at: http://english.mep.gov.cn/down_load/Documents/201104/P020110411532104009882.pdf (14 June 2013).

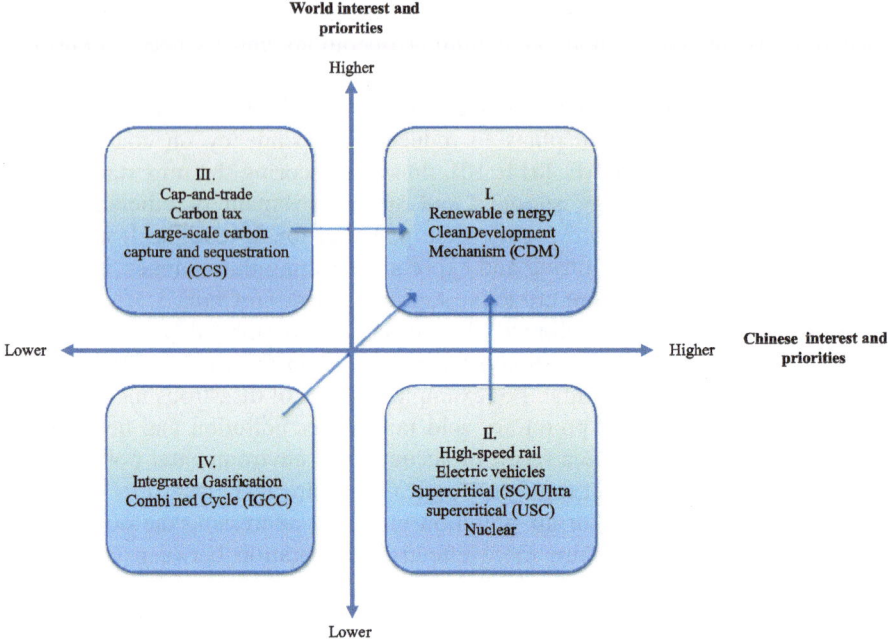

Fig. 2.1 Matrix of world and Chinese interests and priorities

China's 11th FYP heavily depended on command and control approaches, either via target decomposition through the hierarchical governmental levels or through state-owned companies. This top-down system, while effective in setting goals, is inefficient in implementing them. Should China take its targets seriously, it would have to mobilise and allocate its resources to achieve the targets efficiently. The Chinese government has proven in the past that it is capable of achieving a 19.1 percent reduction in energy intensity over the 11th FYP period while maintaining sustained economic growth and rapid urbanisation. Despite these achievements, the system has its disadvantages and a top-down approach has resulted in some unintended consequences. For instance, there were reports of some local authorities shutting down electricity to meet the 20 percent energy intensity reduction goal set by the central government. Brownouts were also reported in the fourth quarter of 2010.[24] Factories, schools and even hospitals were closed on command, even as some manufacturers operated on diesel generators, leading to an increase in the price of diesel. This led further to higher prices of food and other goods that were transported by rail or trucks.

---

[24] "Brownouts Made Energy Efficiency and Emission Reduction" (in Chinese), in: *Xinhua News* (9 September 2010), at: http://news.xinhuanet.com/video/2010-09/11/c_12542322.htm (28 April 2011).

It should be noted that the incentive and accountability system adopted in China might not align with its macro goals, especially when there are conflicting interests between the central and local governments. The central government is likely to have more incentives to protect the environment and climate while local governments may tend to remain more focussed on economic development, based on which their performances are assessed. In instances where local governments receive conflicting targets, they tend to ignore these targets in favour of protecting their own interests such as in terms of increasing revenue, etc. This brings up another issue that further highlights the weaknesses in the Chinese system, namely lack of governance and information integrity. According to reports, even top country leaders acknowledge that some of the performances claimed by local governments are not credible.[25]

Next, the Chinese energy markets generally exist at different levels of reform and liberalisation. For example, China has presently comparatively mature coal markets that are increasingly exposed to global prices. The Chinese power sector, on the other hand, is not market oriented and regulation of the energy system is largely driven by the political priorities of the central government. The conflict that consequently arises between the unevenly reformed coal and power sectors is a key problem for the Chinese energy system: one where the power markets are unable to internalise increased costs. This situation not only further re-enforces the conflict between the coal and power sectors but also adds fuel to the central government's impulse to intervene.

Where renewable energy is concerned, China is faced with the challenge of system integration. For instance, it was estimated that one-third of Chinese wind capacity was not integrated with the grid in 2009 (Li et al. 2012: 787).[26] Although a recent report by the State Grid indicated that the integration rate in China could reach 80 percent,[27] the challenge of system integration and cost still lies in the way of renewable energy development in the country. Currently, China's renewable energy sector is heavily reliant on government subsidies in the form of feed-in-tariffs or direct subsidies, and this has led to overinvestment in certain renewable energy sectors such as wind turbine manufacturing.

---

[25] Batson, Andrew, "Chinese Leader Called Data 'Man-Made'", in: *The Wall Street Journal* (6 December 2010), at: http://online.wsj.com/article/SB10001424052748704156304576003222 932021228.html (8 April 2011).

[26] Liu, Coco, "GRID: China Rebuilds its Power Grid as Part of Its Clean Technologies Push", in: *E&E News* (20 April 2011), at: http://www.eenews.net/public/climatewire/2011/04/20/1 (21 April 2011).

[27] State Grid, "White Paper on Promoting Wind Power Development" (in Chinese) (15 April 2011), at: http://www.sgcc.com.cn/xwzx/gsyw/2011/04/245163.shtml (20 April 2011).

2.6 Future Scenario: Towards a More Market-Oriented Climate Governance

A series of comprehensive economic reform, development and transformation will shape how China and, to a large extent, the international community use energy and address climate change in the coming years. Therefore, a better understanding of China's incentives and constrains will help us to not only comprehend international climate policy but also crystallise possible future policy directions in China. This was indicated in China's 12th FYP, approved by the National People's Congress in March 2011, which aimed to build on the achievements of the 11th FYP with a target energy intensity reduction of 16 percent and, for the first time, a targetted carbon emission intensity reduction of 17 percent based on 2010 levels to be achieved by end 2015 (see Table 2.1).

China is testing a cap-and-trade pilot programme,[28,29] to establish a market in pollution allowances, which is intended to help drive a new 'industrial revolution' of low-carbon technologies. However, this does not necessarily mean that China will accept a physical cap until it reaches peak energy and emission. China did propose a cap in early 2011, suggesting that energy consumption by all sources should not exceed 4 billion tce as of 2015.[30] However, China's total energy consumption in 2010 was 3.25 billion tce, translating to a 4.24 percent growth rate over the years 2011–2015 even as the proposed GDP growth rate for the 12th FYP is at 7 percent. While the figures attached to a cap itself may not be final, the mere notion of a cap delivers a clear message—that an energy limit is not far away and peaking energy demands may be achieved sooner than later. A 'new industry and emerging industries plan,[31] has also recently been unveiled[32] and it is expected to set the tone for China's renewable and clean energy development.

Such initiatives have many takers at the provincial level. For instance, even before the central government had confirmed its endorsement of carbon or

---

[28] "China to Launch Energy Cap-and-trade Trials in Green Push", in: *Reuters News* (5 March 2011), at: http://www.reuters.com/article/2011/03/05/us-china-npc-energy-idUSTRE7240VX2 0110305 (8 April 2011).

[29] NDRC, "Notice on Developing Carbon Cap and Trade Demonstration Program 2011", No. 2601 (29 October 2011), at: http://www.ndrc.gov.cn/zcfb/zcfbtz/2011tz/t20120113_456506.htm (1 December 2012).

[30] "China to Cap Energy Use at 4 Billion Tons of Coal by 2015, Xinhua Reports", in: *Bloomberg News* (4 March 2011), at: http://www.bloomberg.com/news/2011-03-04/china-to-cap-energy-use-at-4-billion-tons-of-coal-by-2015-xinhua-reports.html (28 April 2011).

[31] Energy conservation and environment protection, new generation of information technology, biotechnology, advanced manufacturing, new energy, new materials and new energy vehicles were defined as strategic emerging industries. See, State Council, "12th FYP of National Strategic Emerging Industries", at: http://www.gov.cn/zwgk/2012-07/20/content_2187770.htm (1 December 2012).

[32] State Council, "New Industry and Emerging Industries Plan" (9 July 2012), at: http://www.gov.cn/zwgk/2012-07/20/content_2187770.htm (8 July 2012).

Table 2.1 Comparison of China's 11th and 12th five-year plans

11th FYP (2006–2010)	12th FYP (2011–2015)
"Strong nation, prosperous people"	"Prosperous people, strong nation"
"Ensure continued growth"	"Structural shift"
Economies of scale	Value-add, research and development, innovation
Investment/export	Consumption-driven
"Exclusive growth"	"Inclusive growth"
8 percent GDP	7 percent GDP
Target: 20 percent energy intensity reduction (19.1 percent achieved)	Targets: 16 percent energy intensity reduction; 17 percent carbon intensity reduction
Tertiary sector target: 43 percent (+3 percent)	Tertiary sector target: 50 percent (+7 percent)

GDP gross domestic product
Source NDRC 2011, at: http://www.ndrc.gov.cn/fzgh/ghwb/gjjh/P020110919592208575015.pdf (28 April 2011)
NDRC, "National Economic and Social Development 11th/12th Five-Year Plan" (in Chinese), at: http://www.ndrc.gov.cn/fzgh/ghwb/gjjh/P020110919592208575015.pdf (28 April 2011)

pollution trading, cities such as Beijing, Shanghai and Tianjin were almost simultaneously exploring the feasibility of setting up a climate exchange or environmental exchanges in China. However, before China commits to a cap on carbon emissions, it is more likely that its provinces would first be concerned with building the infrastructure, including the standards, rule of law, policy and other systems for future policy implementation. This shows that China is conscious of the limitations of its command and control policy and is open to introducing more market-oriented approaches to achieve its energy and climate targets. It should be noted that, in the late 1990s and early 2000s, China had tried a cap-and-trade trailer to regulate sulphur dioxide (SO_2) that has not yet been fully adopted.[33] CO_2 is far more complicated than just a pollutant, and the success and failure of climate exchanges will be even harder to chart.

Carbon tax is increasingly seen as a potential instrument for regulating carbon and generating revenue for renewable and clean energy subsidies in China. There are no official statements yet on carbon tax, but several government-related research institutes have undertaken feasibility studies and called for the implementation of a low carbon tax rate.[34] China is currently in the process of short-listing cities where the government could implement and test for the effects of low carbon tax. Through this experiment, China will try to craft and implement market-oriented policy tools to address the weaknesses in its traditional command and control public administrative structure. Comprehensive climate governance through such an approach would mean combining administrative and market-oriented tools (Qi et al. 2007: 851; Zhang 2008: 81).

---

[33] CEC, "Difficulties Cloud Cap and Trade Pilots", in: *Economic Reference News* (8 November 2011), at: http://www.cec.org.cn/xinwenpingxi/2011-11-08/73955.html (8 April 2012).
[34] "China to Introduce Carbon Tax: Official", in: *Xinhua News* (19 February 2013), at: http://news.xinhuanet.com/english/china/2013-02/19/c_132178898.htm (20 February 2013).

2.7 Conclusion and Discussion: Implications for International Energy and Climate Policy

China's development in clean energy has attracted global attention. It is proactively pursuing clean energy technology at a time when there appears to remain unresolved conflicts between developed and developing economies within the global climate change regime. Amid the impasse in the international climate change regime, China is testing more market-oriented governance in its 12th FYP and beyond. The new targets and their implementation should be watched closely, as they may influence the future shape of the global energy and climate policy maps. More importantly, the Chinese experience offers a unique perspective on how other emerging and developing economies could implement climate change policies at the national level.

China is undergoing major shifts in its policies and the government views climate change as an avenue for economic growth. This has been made possible by strategic shifts in both leadership and political thinking, which has been reinforced by a shift in public consciousness. The government in China as a result appears to be more open, flexible and proactive. Besides this, the institution for confronting climate challenge has been strengthened further as a result of the realisation that climate change policy in China cannot be crafted independent of the country's economic, environment and climate security. There is growing support for the importance of dealing with climate change within the energy-climate-water nexus and energy-climate-food nexus.

What does all this mean for international climate change policy? As Morse et al. argue, China's pursuit for energy security, economic growth and macro-economic stability will work against large-scale CCS in the country unless the latter is strongly supported by external funding (Morse et al. 2009: 5). Therefore, if the international community could align its policies and invest in China's clean technology, the collective benefit of a reduction in China's pollution would be enjoyed by all. Apart from this, investment in China's clean technology would also lead to technological transfers and exports to the rest of the world.

The lessons He and Morse draw from the Chinese wind Clean Development Mechanism (CDM) experience show how an international climate policy can be ineffective if it does not consider China's national, political, economic and domestic constraints (He/Morse 2010: 2). This experience alone makes for a compelling argument for greater international cooperation rather than competition with China. Moving forward, the international community should identify and understand opportunities and areas of cooperation that they could possibly leverage with China.

Acknowledgments The paper was presented at a conference on *Policy Responses to Climate Change and Energy Security Post-Cancún: Implications for the Asia–Pacific Region's Energy Security*, held in March 2011 in Singapore, that was organised by the Energy Studies Institute (ESI) at the National University of Singapore (NUS). The author would like to thank Chou Siaw Kiang, Hooman Peimani, Nur Azha Putra and Jan Lui of ESI for the invitation and discussion, as well as Daniel Kammen, David Victor, Richard Morse, Joe Chang, Mark Thurber and Frank Wolak for their insights and support. The views reflected in the paper are the author's sole responsibility.

References

CAE (Chinese Academy of Engineering), 2011: *China's Medium and Long-Term Energy Development Strategy* (Beijing: Science Press).

Figueres, Christiana, 2010: "Address by Christiana Figueres, Executive Secretary United Nations Framework Convention on Climate Change", Ministerial Consultation on Climate Change and the Cancun Conferences, 25 September 2010, New York (New York: UNFCCC): 3.

FON (Friends of Nature), 2011: *The 2010 Annual Report on China's Environment* (Beijing: Social Science Press).

Fridley, D.; Zheng, N.; Zhou, N.; Ke, J.; Hasanbeigi, A.; Morrow, B.; Price, L., 2011: *China Energy and Emissions Path to 2030* (Berkeley: Lawrence Berkeley National Liboratory).

Giddens, A., 2009: *The Politics of Climate Change* (Cambridge, UK: Polity Press).

Hale, Thomas; Held, David; Young, Kevin, 2013: *Gridlock: Why Global Cooperation Is Failing When We Need It Most* (Cambridge, UK: Polity Press).

He, G., 2007: "Chinese Society Confronted with Climate Change", in: *China Perspectives*, 1: 77–82.

He, Gang; Morse, Richard K., 2010: "Making Carbon Offsets Work in Developing World: Lessons from Chinese Wind Controversy", PESD Working Paper #90, March (Stanford: Program on Energy and Sustainable Development, 2010).

IEA (International Energy Agency), 2010: *World Energy Outlook 2010* (Paris: OECD/IEA).

Jiang, K.; Hu, X.; Liu, Q.; Zhuang, X.; Liu, H., 2010: "2050 China Low Carbon Development Scenario Research", in: 2050 CEACER (Ed.): 2050 *China Energy and CO_2 Emissions Report* (Beijing: Science Press).

Keohane, R.O.; Victor, D.G., 2011: "The Regime Complex for Climate Change", in: *Perspectives on Politics*, 9: 7–23.

Li, J.; Cai, F.; Qiao, L.; Xie, H.; Gao, H.; Yang, X.; Tang, W.; Wang, W.; Li, X., 2012: *China Wind Power Outlook 2012* (Beijing: China Environment Science Press).

Lieberthal, K.; Sandalow, D., 2009: *Overcoming Obstacles to US–China Cooperation on Climate Change* (Washington, D.C.: Brookings Institution).

MOST (Ministry of Science and Technology of the People's Republic of China); CAS (Chinese Academy of Science); CMA (China Meteorological Administration), 2007: *National Assessment Report on Climate Change* (Beijing: Science Press).

MOST (Ministry of Science and Technology of the People's Republic of China); CMA (China Meteorological Administration); CAS (Chinese Academy of Science), 2011: *National Assessment Report on Climate Change* (Beijing: Science Press).

Morse, Richard K.; Rai, Varun; He, Gang, 2009: "Real Drivers of Carbon Capture and Storage in China and Implications for Climate Policy", PESD Working Paper #88, August (Stanford: Program on Energy and Sustainable Development, 2009).

NBS (National Bureau of Statistics of China), 2010: *China Energy Statistical Yearbook 2009* (Beijing: National Statistics Press).

Peng, Wuyuan, 2009: "The Evolution of China's Coal Institutions", PESD Working Paper Series #86, August (Stanford: Program on Energy and Sustainable Development, 2009).

Price, Lynn; Levine, Mark D.; Zhou, Nan; Fridley, David; Aden, Nathaniel; Lu, Hongyou; McNeil, Michael; Zheng, Nina; Qin, Yining; Yowargana, Ping, 2011: "Assessment of China's Energy-saving and Emission-reduction Accomplishments and Opportunities During the 11th Five Year Plan", in: *Energy Policy*, 39,4: 2165–2178.

Qi, Y.; Ma, L.; Zhang, L., 2007: "Climate Change Governance in China: A Case Study", in: *China Population, Resources and Environment*, 17: 8–12.

Rui, Huaichuan; Morse, Richard K.; He, Gang, 2010: "Remaking the World's Largest Coal Market: The Quest to Develop Large Coal-power Bases in China", PESD Working Paper #98, December (Stanford: Program on Energy and Sustainable Development, 2010).

Victor, D.G., 2011: *Global Warming Gridlock: Creating More Effective Strategies for Protecting the Planet* (Cambridge, UK: Cambridge University Press).

Wang, J., 2011: "China's Search for a Grand Strategy—A Rising Great Power Finds Its Way",
  in: *Foreign Affairs*, 90: 68.
Wang, T.; Watson, J., 2010: "Scenario Analysis of China's Emissions Pathways in the 21st
  Century for Low Carbon Transition" in: *Energy Policy*, 38: 3537–3546.
Zhang, H., 2006: "China's Position in the Negotiations on International Climate Change:
  Continuities and Changes", in: *International Policy*, 12: 276–314.
Zhang, S., 2008: "China: Facing the Challenges to Link Climate Change Responses with
  Sustainable Development and Local Environmental Protection", in: *Globalizations*, 5: 81–82.
Zhou, N.; Fridley, D.; McNeil, M.; Zheng, N.; Ke, J.; Levine, M., 2011: *China's Energy and
  Carbon Emissions Outlook to 2050* (Berkeley: Lawrence Berkeley National Laboratory).

Abbreviations

CAE	Chinese Academy of Engineering
CAS	Chinese Academy of Science
CCS	Carbon capture and sequestration
CCUS	Carbon Capture, Utilisation and Sequestration
CDM	Clean Development Mechanism
CEC	China Electricity Council
CMA	China Meteorological Administration
CO_2	Carbon dioxide
COD	Chemical oxygen demand
ERG	Energy and Resources Group
FON	Friends of Nature
FYP	Five-Year Plan
GDP	Gross domestic product
GW	Gigawatt
IGCC	Integrated Gasification Combined Cycle
MEP	Ministry of Environmental Protection of the People's Republic of China
MOST	Ministry of Science and Technology of the People's Republic of China
NBS	National Bureau of Statistics of China
NDRC	National Development and Reform Commission
OECD	Organisation for Economic Co-operation and Development
PESD	Program on Energy and Sustainable Development
SC/USC	Supercritical/Ultra supercritical
SO_2	Sulphur dioxide
tce	Tonnes of coal equivalent
US	Unites States of America
WRI	World Resources Institute

Chapter 3
India's Efforts to Maintain and Enhance Energy Security While Reducing Greenhouse Gas Emissions

Harbans L. Bajaj

Abstract India faces the twin challenges of sustaining its fast economic growth, which is essential for improving the quality of life of its 1.21 billion people, and adapting to climate change. India has an abundance of coal and strong potential in hydroenergy, but limited oil and gas reserves. Coal will thus continue to play a major role in meeting India's growing energy demands. To reduce greenhouse gas emissions, the government will focus on developing the country's renewable energy sources such as hydro and nuclear energy, deploying clean coal technologies, as well as improving energy conservation and enhancing energy efficiency. To this end, the Prime Minister's Council on Climate Change has inter alia suggested three key national missions namely, National Solar Mission, National Mission for Enhanced Energy Efficiency and National Mission for a "Green India".

Keywords Clean development mechanism · Climate change · Coal beneficiation · Energy conservation · Energy efficiency · Environmentally benign policies · Greenhouse gas emissions · Renewable energy sources · Supercritical technology

3.1 Introduction

The population of India, which grew to 1.21 billion in 2011, has recorded a growth of 17.5 percent in the last 10 years (MIB 2013: 7). Given its rapidly expanding population and economy, India will require considerably more energy to meet its growing energy demands. Apart from its massive population, India faces the additional challenges of sustaining its fast economic growth, which imposes considerable primary energy demand, and dealing with the global threat of climate change.

H. L. Bajaj (✉)
S-451, Greater Kailash Part-II, New Delhi 110048, India
e-mail: BAJAJHL@gmail.com

N. A. Putra and E. Han (eds.), *Governments' Responses to Climate Change:*
Selected Examples From Asia Pacific 10, SpringerBriefs in Environment, Security,
Development and Peace, DOI: 10.1007/978-981-4451-12-3_3, © The Author(s) 2014

India is the world's fourth largest consumer of energy, with a total primary energy demand of 621 million tonnes of oil equivalent (mtoe) in 2010 (IEA 2010: 605). India's average annual gross domestic product (GDP) grew at 7.3 percent between 2000 and 2008 whereas primary energy demand increased at 3.8 percent per annum (IEA 2010: 605–606). However, in spite of reporting such impressive figures, some 400 million people in India were without access to electricity (IEA 2010: 606); more than twice that number rely on the traditional use of biomass for cooking (IEA 2010: 605–606). Also, despite being the fifth largest producer of electricity after China, the United States, Japan and Russia, India's annual per capita consumption of electricity at about 879 kilowatt hour[1] (kWh) is only about 30 percent of the global average. In order to attain even the world's average level of per capita consumption of electricity, India will need to triple its present-generation installed capacity.

Coal, being available in abundance, is the major source of energy in India and contributes to about 68 percent of its electricity demand.[2] Coal is expected to grow by some 3.3 percent per annum to 2030.[3] However, the scarcity of indigenous hydrocarbon resources has led India to import a growing share of its energy. Even with extensive steam coal reserves, India imports coal to meet domestic consumption requirements. Expanding energy infrastructure in the oil and gas and electricity sectors is a major priority for the country.

Even as India is attempting to improve its vital statistics on the economy front, climate change has evolved as a formidable and complex challenge needing close cooperation among all nations. Activities such as energy generation from fossil fuels, industrialisation and deforestation have been increasing the atmospheric concentration of greenhouse gases (GHGs) beyond their natural levels, resulting in global climate change. Global warming can be mainly attributed to carbon dioxide (CO_2), which is a major contributor to the greenhouse effect. In most countries, the major requirement of power is met through thermal power plants. India too depends largely on coal as a major source of energy for producing power, coal is expected to continue to play a major role in producing power for the country in the near future. Therefore, thermal power plants, being potential means of mitigating GHG emissions, need to be managed efficiently to keep the generation of CO_2 at optimally low values.

As the challenges that India faces in meeting its increasing energy requirements while mitigating the threats of climate change are grave, it is endeavouring to ensure its energy security by adopting policies that are also conducive to mitigating the effects of GHG emissions. Main areas of thrust are increasing efficiencies, aggressively deploying renewable energy sources, harnessing hydropower,

---

[1] CEA, "Growth of Electricity Sector in India from 1947-2012", June 2012, at: http://www.npti.in/Download/Misc/CEA%20Growth%20of%20Electricity.pdf (20 June 2013): 6.

[2] WEC, "2010 Survey of Energy Resources", at: http://www.worldenergy.org/documents/ser_2010_report_1.pdf (18 July 2013): 4.

[3] Ibid.

adopting state-of-the-art clean technologies, conserving energy, retiring older inefficient generating plants, formulating new policies that encourage competition, improving efficiency and preventing the pilferage of electricity, and increasing its green cover.

3.1.1 Climate Change and the Kyoto Protocol

India is a signatory to the Kyoto Protocol, which is making substantial global efforts to mitigate GHG emissions. Post-Cancún, as per the note FCCC/AW-GLCA/2011/INF.1 stipulated on 18 March 2011 and issued by the Secretariat of Ad Hoc Working Group on Long-term Cooperative Action under the Convention (AWG-LCA), United Nations Framework Convention on Climate Change (UN-FCCC),[4] India has proposed the following Nationally Appropriate Mitigation Actions (NAMAs):

> 71. India communicated that it will endeavour to reduce the emissions intensity of its GDP by 20–25 percent by 2020 compared with the 2005 levels. It added that emissions from the agriculture sector would not form part of the assessment of its emissions intensity.

> 72. India stated that the proposed domestic actions are voluntary in nature and will not have a legally binding character. It added that these actions will be implemented in accordance with the provisions of relevant national legislation and policies, as well as the principles and provisions of the Convention, in particular Article 4, paragraph 7. Finally, it added that this information has been communicated in accordance with the provisions of Article 12, paragraphs 1(b) and 4, and Article 10, paragraph 2(a), of the Convention.[5]

3.1.2 India's Carbon Dioxide Emissions

India's per capita CO_2 emissions from the consumption of energy in 2010 were 1.365 metric tons of CO_2 per person, which is among the lowest in the world.[6] The average per capita CO_2 emissions were 18.222 metric tons of CO_2 per person for

---

[4] UNFCCC, "Compilation of Information on Nationally Appropriate Mitigation Actions to be Implemented by Parties Not Included in Annex I to the Convention", FCCC/AWGLCA/2011/INF.1, 18 March 2011, Note issued by the secretariat of Ad Hoc Working Group on Long-term Cooperative Action under the Convention, at: http://unfccc.int/resource/docs/2011/awglca14/eng/inf01.pdf (18 June 2013).

[5] Ibid: 26.

[6] EIA, "Indicators: Asia and Oceania, India: Per Capita Carbon Dioxide Emissions from the Consumption of Energy (Metric Tons of Carbon Dioxide per Person)", in: *International Energy Statistics*, at: http://www.eia.gov/cfapps/ipdbproject/iedindex3.cfm?tid=90&pid=45&aid=8&cid=r7,IN,&syid=2007&eyid=2011&unit=MMTCD (16 July 2013).

Table 3.1 India's annual absolute CO_2 emissions (from 2006–2007 to 2011–2012)

	2006–2007	2007–2008	2008–2009	2009–2010	2010–2011	2011–2012
Absolute CO_2 emission (million tonnes)	494	520	548	580	598	637

Source CEA 2012, at: http://cea.nic.in/reports/planning/cdm_co2/cdm_co2.htm (16 July 2013)
CEA 2013, at: http://www.cea.nic.in/reports/planning/cdm_co2/cdm_co2.htm (16 July 2013)
CEA, "Baseline Carbon Dioxide Emission Database Version 7.0", at: http://cea.nic.in/reports/planning/cdm_co2/cdm_co2.htm (16 July 2013)
CEA, "Baseline Carbon Dioxide Emission Database Version 8.0", at: http://www.cea.nic.in/reports/planning/cdm_co2/cdm_co2.htm (16 July 2013)
CO_2 = carbon dioxide

the United States[7] and 6.012 metric tons for China[8] for the same year. According to data compiled by India's Central Electricity Authority (CEA) on CO_2 emissions from the Indian power sector, which was based on data furnished by its power stations, an estimated 597 million tonnes of CO_2 was emitted in 2010–2011 from grid-connected power stations in India.[9] Notably, the average emissions from thermal generation have reduced from 1.01 kg of CO_2/kWh in the year 2008–2009 to 0.97 kg of CO_2/kWh in the year 2011–2012.[10] Table 3.1 shows the annual per capita CO_2 emissions from the Indian power sector during the last 6 years.

This chapter provides an exhaustive overview of the Indian energy sector, particularly electricity, and discusses the various steps being taken by the government to ensure energy security while focussing on reducing CO_2 emissions.

3.2 The Indian Power Sector: An Overview

India's per capita electricity consumption, which was 16 kWh in the year 1947,[11] increased to 879 kWh during 2011–2012.[12] The National Electricity Policy (NEP) report of the Government of India (GoI) stipulated that per capita electricity

---

[7] EIA, "Indicators: North America: Per Capita Carbon Dioxide Emissions from the Consumption of Energy (Metric Tons of Carbon Dioxide per Person)", in: *International Energy Statistics*, at: http://www.eia.gov/cfapps/ipdbproject/iedindex3.cfm?tid=90&pid=45&aid=8&cid=r1,&syid=2007&eyid=2011&unit=MTCDPP (18 July 2013).

[8] EIA, "Indicators: Asia & Oceania: Per Capita Carbon Dioxide Emissions from the Consumption of Energy (Metric Tons of Carbon Dioxide per Person)", in: *International Energy Statistics*, at: http://www.eia.gov/cfapps/ipdbproject/iedindex3.cfm?tid=90&pid=45&aid=8&cid=r7,&syid=2007&eyid=2011&unit=MTCDPP (18 July 2013).

[9] CEA, "Baseline Carbon Dioxide Emission Database Version 7.0", at: http://cea.nic.in/reports/planning/cdm_co2/cdm_co2.htm (16 July 2013).

[10] CEA, "Baseline Carbon Dioxide Emission Database Version 8.0", at: http://www.cea.nic.in/reports/planning/cdm_co2/cdm_co2.htm (16 July 2013).

[11] CEA, "Growth of Electricity Sector in India from 1947-2012", June 2012, at: http://www.npti.in/Download/Misc/CEA%20Growth%20of%20Electricity.pdf (20 June 2013): 6.

[12] Ibid: 5.

Table 3.2 India's installed generating capacity (in MW) as on 31 December 2012

Mode	Installed capacity (MW)
Thermal	140,976
Coal	120,873
Gas	18,903
Diesel	1,200
Hydro	39,339
Nuclear	4,780
Renewable	25,856
Total	210,951

Source CEA 2012, at: http://www.cea.nic.in/reports/monthly/inst_capacity/dec12.pdf (13 July 2013)

CEA, "Monthly Review of Power Sector Performance", December 2012, at: http://www.cea.nic.in/reports/monthly/inst_capacity/dec12.pdf (13 July 2013)

MW = megawatt

consumption would be over 1,000 kWh by 2012.[13] According to the 18th Electric Power Survey of India (EPS) conducted by the CEA, electrical energy requirement and annual peak electric load in India would be 1,354.874 gigawatt hour (GWh) and 199.540 gigawatts (GW), respectively, for the 2016–2017 period (CEA 2010: 31).

India's installed generating capacity at the end of December 2012 was around 210,951 megawatts (MW) (Table 3.2).[14] Figure 3.1, which details its installed capacity in terms of sources, indicates that coal was the primary source of power generated in 2012. Similarly, Fig. 3.2, which gives the sectorwise break-up of installed capacity in end 2012, shows that state and central actors continued to be major participants in the power sector.

Interestingly, private sector participation in the power sector increased from 12.6 percent in March 2007 (CEA 2008: 32) to 29 percent in December 2012,[15] as a result of reforms in the sector. Now, generation does not require licence, and open access in transmission and distribution (T&D) has been allowed to foster competition. Electricity regulators are in place too, both in the states and at the centre.

Figure 3.3 provides a break-up of the total power generation targeted for the period 2012–2013.

As these figures show, total thermal-based generation that was targeted during the 2012–2013 period was 82 percent of the total generation (Fig. 3.3) although only around 66 percent of installed capacity was from thermal power sources

[13] CEA, "National Electricity Policy" (12 February 2005), at: http://powermin.nic.in/whats_new/national_electricity_policy.htm (13 July 2013).

[14] CEA, "Monthly Review of Power Sector Performance", December 2012, at: http://www.cea.nic.in/reports/monthly/inst_capacity/dec12.pdf (13 July 2013).

[15] Based on data computations. See also CEA, "Monthly Review of Power Sector Performance", December 2012, at: http://cea.nic.in/reports/monthly/inst_capacity/dec12.pdf (13 July 2013): 15.

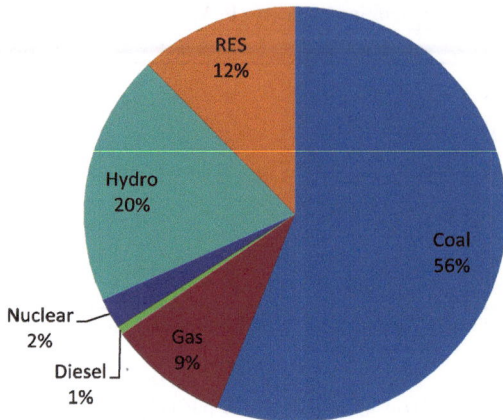

Fig. 3.1 Break-up of all-India installed capacity (in MW) based on source type as on 31 March 2012. *Source* CEA 2012. CEA, "Monthly Review of Power Sector Performance", March 2012, at: http://www.cea.nic.in/reports/monthly/inst_capacity/mar12.pdf (13 July 2013), at: http://www.cea.nic.in/reports/monthly/inst_capacity/mar12.pdf (13 July 2013). Total installed capacity = 1,99,877.03 MW. MW Megawatt; RES Renewable energy source

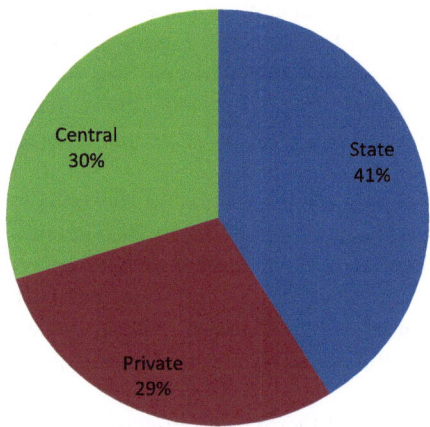

Fig. 3.2 Sectorwise share of all-India installed capacity as on 30 November 2012. *Source* CEA 2012. CEA, "Monthly Review of Power Sector Performance", December 2012, at: http://www.cea.nic.in/reports/monthly/inst_capacity/dec12.pdf (13 July 2013), at: http://www.cea.nic.in/reports/monthly/inst_capacity/dec12.pdf (13 July 2013). Total installed capacity = 210936.71 MW. MW Megawatt

(including coal (56 percent), gas (9 percent) and diesel (1 percent)) in March 2012 (Fig. 3.1). In other words, the bulk of power generation in India is from sources that contribute to CO_2 emissions.

In spite of massive additions to generation and T&D capacities over the last 60 years, growth in demand for power has mostly exceeded the augmentation in India's generation capacity. As a result, peak power deficit and energy shortages of

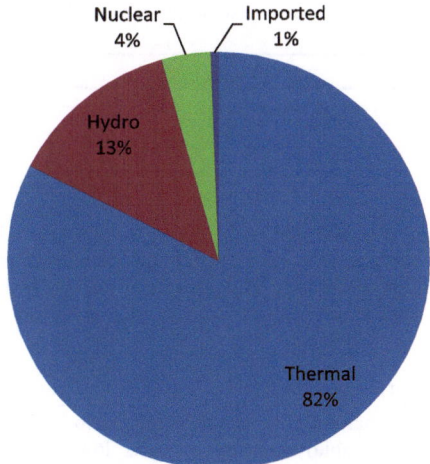

Fig. 3.3 Target generationfor 2012–2013. *Source* CEA 2013. CEA, "Monthly Review of Power Sector (Executive Summary)", March 2013, at: http://cea.nic.in/reports/monthly/executive_rep/ mar13/mar13.pdf (13 July 2013): 1: 1, at: http://cea.nic.in/reports/monthly/executive_rep/mar13/ mar13.pdf (13 July 2013). Total generation = 930,000 MU, excluding renewable energy

varying magnitudes are often experienced in the country. For instance, during April 2011–March 2012, India faced an energy shortage of around 8.5 percent and a peak deficit of 10.6 percent.[16] According to the 18th EPS, peak demand by 2017 would be 199 GW while energy requirement would be 1,355 billion kWh (CEA 2010: 31). To meet this energy demand, India will need not only huge capacity addition but also to build a corresponding transmission system.

3.2.1 Capacity Addition Programmes for the 12th (2012–2017) and 13th (2017–2022) Five-Year Plans

According to the CEA, a capacity addition of nearly 80,000 MW from conventional sources of energy (that is from thermal, hydro and nuclear sources) would be required during the 12th Five-Year Plan (FYP; 2012–2017) (Table 3.3) in addition to 30,000 MW capacity from renewable energy sources during the same period.[17] However, under the base case scenario, capacity addition of only 18,500 MW from renewable energy sources would be required for the 12th FYP.[18]

---

[16] CEA, "Growth of Electricity Sector in India from 1947–2012", June 2012, at: http:// www.npti.in/Download/Misc/CEA%20Growth%20of%20Electricity.pdf (20 June 2013): 64.

[17] CEA, "National Electricity Plan: Generation", January 2012, at: http://www.cea.nic.in/ reports/powersystems/nep2012/generation_12.pdf (18 June 2013): 239.

[18] Ibid.

Table 3.3 Capacity addition programme (excluding renewable) for the 12th Five-Year Plan (2012–2017)

Source	Total (MW)
Hydro	9,204
Thermal	67,686
Nuclear	2,800
Total (excluding renewable)	79,690

Source CEA 2012: CEA, "National Electricity Plan: Generation", January 2012, at: http://www.cea.nic.in/reports/powersystems/nep2012/generation_12.pdf (18 June 2013): 239; at: http://www.cea.nic.in/reports/powersystems/nep2012/generation_12.pdf (18 June 2013)
MW = megawatt

Table 3.4 Estimated capacity requirement for the 13th Five-Year Plan (2017–2022)

Source	Scenario-1 (low gas, low renewable) (base case)	Scenario-2 (high gas, low renewable)	Scenario-3 (high gas, high renewable)
Hydro	12,000	12,000	12,000
Thermal			
Gas	0	13,000	13,000
Coal	49,200	38,000	34,000
Subcritical	10,000	0	0
Supercritical	39,200	38,000	34,000
Nuclear	18,000	18,000	18,000
Total capacity planned from conventional sources	79,200	81,000	77,000
Renewable			
Excluding solar	14,500	14,500	25,000
Solar	16,000	16,000	20,000

Source CEA 2012: 110–111; CEA, "National Electricity Plan: Generation", January 2012, at: http://www.cea.nic.in/reports/powersystems/nep2012/generation_12.pdf (18 June 2013): 239, at: http://www.cea.nic.in/reports/powersystems/nep2012/generation_12.pdf (18 June 2013)
MW = megawatt

The CEA has projected various likely scenarios of power demand for the 13th FYP,[19] as summarised in Table 3.4.

3.2.2 Rural Electrification

According to the CEA, in December 1950, only 3,061 villages in the country had access to electricity.[20] The electrification scenario has changed significantly since

---

[19] Ibid: 110–111.

[20] CEA, "Growth of Electricity Sector in India from 1947–2012", June 2012, at: http://www.npti.in/Download/Misc/CEA%20Growth%20of%20Electricity.pdf (20 June 2013): 5.

then, as 557,439 villages in India have been electrified as of March 2012.[21] Many of the unelectrified villages are located in rural areas and it is not possible to extend power supply to these villages through the existing power grid. The Indian government has launched an ambitious scheme called the Rajiv Gandhi Grameen Vidyutikaran Yojana (RGGVY),[22] which is a rural electrification scheme that aims to electrify all unelectrified villages in India.

3.2.3 Reforms and Restructuring

The focus of power sector reforms in India has been the improvement of the quality of power supply, ensuring the metering of all consumers and energy audits, improvement of billing and collection efficiency, reduction of technical and commercial losses, reduction and elimination of theft of power, constitution and operationalisation of State Electricity Regulatory Commissions (SERCs), organisational restructuring to focus on accountability for performance, and achieving commercial viability in a time-bound manner.

India enacted the Electricity Act 2003[23] (the Act) to create an environment for enhanced competition, promote efficient and environmentally benign policies, and encourage private sector participation. As per the Act, SERCs are *inter alia* required to promote the cogeneration and generation of electricity from renewable sources of energy by providing suitable measures for connectivity with the grid for sale of electricity to any person, and also to specify for the purchase of electricity from renewable sources a percentage of the total consumption of electricity in the area of a distribution licensee. Provisions under the Act, effective from 10 June 2003, have been made more stringent against the theft of electricity. Recognising that curbing the theft of electricity could help prevent the misuse and inefficient use of energy, the Act also provides for the constitution of Special Courts for the purpose of providing speedy trials of offences related to such theft.

3.2.4 National Electricity Policy

The GoI has notified the NEP, which *inter alia* recommends the promotion of unconventional energy sources.[24] Keeping in view the availability of energy

---

[21] Ibid.

[22] MoP, "Rajiv Gandhi Grameen Vidyutikaran Yojana: Scheme for Rural Electricity Infrastructure and Household Electrification", at: http://rggvy.gov.in/rggvy/rggvyportal/index.html (28 July 2013).

[23] MoP, "Electricity Act, 2003", at: http://www.powermin.nic.in/acts_notification/electricity_act2003/preliminary.htm (18 June 2013).

[24] CEA, "National Electricity Policy" (12 February 2005), at: http://powermin.nic.in/whats_new/national_electricity_policy.htm (13 July 2013).

resources, the technology available for the exploitation of these resources, the economics of generation using different resources and energy security issues, the NEP aims at laying guidelines for accelerated development of the power sector, providing electricity supply to all areas, and protecting the interests of consumers and other stakeholders. The policy aims at providing electricity to all and increasing per capita electricity availability to over 1,000 units by 2012.[25]

3.2.5 National Electricity Plan

The CEA is mandated to frame the National Electricity Plan every 5 years and revisit it from time to time in accordance with the NEP. The Plan serves as a road map toward optimum growth of the power sector. The CEA formulates short-term and prospective plans for the development of the electricity system and coordinates the activities of various planning agencies for the optimal utilisation of resources to serve the interests of the national economy.

3.3 Coal-Based Thermal Generation and the Indian Power Sector: Technological Developments

Rapid technological development has been witnessed in the Indian power sector over the last few decades. The largest unit size of coal-based plants, which was a mere 30 MW in the 1950s, rapidly increased to 60 MW in the 1960s and to 110/120/140 MW in the 1970s.[26] Next, 200 MW units of Russian Leningradsky Metallichesky Zavod (LMZ) and Siemens KWU (Siemens Power Generation Group) designs were introduced in 1977 and 1983, respectively.[27] Then, 500 MW units were introduced in 1984.[28] Four supercritical units of 660 MW have been commissioned until November 2011.[29] Currently, only a handful of 800 MW units are in operation.[30] As indicated in the Table 3.5, the increase in unit size has been associated with corresponding increase in steam parameters such as pressure, temperature and efficiency levels.

With rapidly expanding thermal generation capacity, the installation of large-size supercritical pressure coal-based thermal units is being encouraged in India to

---

[25] Ibid.

[26] CEA, "National Electricity Plan: Generation", January 2012, at: www.cea.nic.in/reports/powersystems/nep2012/generation_12.pdf (18 June 2013): 46.

[27] Ibid.

[28] Ibid.

[29] Ibid: 47.

[30] Ibid: 46.

Table 3.5 Various unit sizes and main parameters

Unit size (MW)	MS pressure (kg/ cm^2)	MS/RH temperature (°C)	Gross design efficiency (%)
30–50	60	482	28.2
60–100	90	535	31.3
210 LMZ	130	535/535	35.63
210 KWU	150	535/535	37.04
250	150	535/535	38.3
500	169	538/538	38.6*
660 (supercritical)	247	535/565	39.5
		565/593	40.5
800	247	565/593	40.5

Source CEA 2012: CEA, "National Electricity Plan: Generation", January 2012, at: http://www.cea.nic.in/reports/powersystems/nep2012/generation_12.pdf (18 June 2013): 46, at: http://www.cea.nic.in/reports/powersystems/nep2012/generation_12.pdf (18 June 2013)
°C = degree Celsius; cm = centimetre; kg = kilogramme; KWU = Siemens Power Generation Group; LMZ = Leningradsky Metallichesky Zavod; MS = main steam; MW = megawatt; RH = reheat
* With turbine-driven boiler feed pump

achieve enhanced efficiency, reduced coal consumption and GHG emissions, and faster capacity additions. The steam parameters being adopted for these units are 247 kg/cm^2, 535/565 °C.[31] Supercritical units of 660/800 MW capacity are also being planned in a big way under the Ultra Mega Power Projects (UMPP) scheme.[32]

3.3.1 Efficiency of Coal-Based Thermal Generation

As of March 2011, there were 107 steam power stations in India, with an aggregate installed capacity of 93,513 MW (CEA 2012: 111). Out of these 107 steam stations, 95 were coal-fired, 10 were lignite-fired and two were multifuel-fired steam turbine power stations (CEA 2012: 111). Among the above 107 steam stations, reported overall thermal efficiency in 2010–2011 was 25 percent for 18 stations (aggregate installed capacity, 5,739 MW), between 25 and 30 percent for 24 stations (aggregate installed capacity, 17,452 MW) and above 30 percent for 65 stations (aggregate installed capacity, 70,322 MW) (CEA 2012: 111).

At present, a majority of coal-fired units in India are subcritical pulverised coal units. The size of these units ranges from 60 to 660 MW, with 200/210/250 MW units forming a majority of coal/lignite-based installed capacity (CEA 2012: 111). While the efficiency of higher-sized coal units is in the range of 34–37 percent on

---

[31] Ibid.

[32] Ibid: 47.

Table 3.6 All-India thermal efficiency of coal/lignite-based plants from 2004–2005 to 2010–2011

Years	Thermal efficiency (%)
2004–2005	32.16
2005–2006	32.73
2006–2007	32.44
2007–2008	32.69
2008–2009	32.70
2009–2010	32.53
2010–2011	32.73

Source CEA (2012: 115)

gross calorific value (GCV) basis, older units are running at much lower efficiencies (CEA 2012: 111). The overall thermal efficiency of coal/lignite-based units, including older units, is estimated to be in the range of 33 percent (CEA 2012: 111). Table 3.6 provides the all-India thermal efficiency of coal/lignite-based plants from 2004–2005 to 2010–2011 (CEA 2012: 115).

3.4 Low-Carbon Growth for India: Strategies and Initiatives

India has decided to adopt a low-carbon growth strategy to reduce its energy intensity and promote sustainable development. As defined by the Brundtland Commission, sustainable development is one that "meets the needs of the present without compromising the ability of future generations to meet their own needs" (WCED 1987: 43). In line with this definition and with a view to reduce GHG emissions, the key approaches being pursued in India include the harnessing of renewable resources to the extent possible (non-conventional sources), promotion of hydro and nuclear generation (conventional sources), enhancing the efficiency of existing power plants, and the introduction of new technologies for power generation for enhancing efficiency, better demand-side management (DSM) and conservation. Since coal will continue to dominate power generation in the near future, supercritical technology has been proposed as the way forward to further enhance the efficiency of coal-fired thermal generation. To produce power, supercritical technology will convert coal far more efficiently than pulverised coal boiler technology with subcritical steam parameters. Initiatives have also been

---

[33] Integrated Gasification Combined Cycle (IGCC) technology enables the efficient use of coal by its gasification as a clean fuel, which is used in a gas turbine for power generation. The flue gases from the gas turbine are lead to a waste heat recovery steam generator. Steam thus produced is fed into a steam turbine- generator set producing electricity, thereby resulting in increased efficiency and reduced greenhouse gas (GHG) emissions.

taken to undertake feasibility studies for the introduction of integrated gasification combined cycle[33] (IGCC) technology for power generation.

3.4.1 Clean Development Mechanism Projects

Clean Development Mechanism (CDM) is a mechanism created under the Kyoto Protocol that allows a country with an emission-reduction or emission-limitation commitment under the Protocol to implement an emission-reduction project in developing countries.[34] The Ministry of Environment and Forests (MoEF), is the nodal agency for dealing with issues related to CDM in India. The National CDM Authority (NCDMA), which has interministerial participation, is the Designated National Authority (DNA) that accords host country approvals for Indian CDM projects, as this is one of the prerequisites for such projects. Of the total projects submitted in the world, 1,317 projects where India is the host country have been registered by the CDM Executive Board.[35]

The CO_2 baseline emission data published by the CEA for the Indian power sector helps to promote CDM projects aimed at mitigating climate change by aiding CDM project developers reduce their transaction costs. Several projects that are based on the CEA's baseline data have already been registered by the CDM Executive Board.

3.4.2 Introduction of New Technologies: Clean Coal Technologies

In accordance with India's low-carbon growth strategy, the adoption of clean coal technologies, including the addition of supercritical units, and the promotion of IGCC and circulating fluidised bed combustion (CFBC) technologies are being undertaken in future FYPs in a big way. For instance, in the 12th FYP, nearly 50 percent of coal-based capacity is being planned on supercritical units whereas, in the 13th FYP, all new coal-based capacity is likely to be based on supercritical technology.[36]

---

[34] UNFCCC, "Clean Development Mechanism (CDM)", at: http://unfccc.int/kyoto_protocol/ mechanisms/clean_development_mechanism/items/2718.php (18 June 2013).

[35] CDM, "Project Search", at: http://cdm.unfccc.int/Projects/projsearch.html (16 July 2013).

[36] CEA, "National Electricity Plan: Generation", January 2012, at: www.cea.nic.in/reports/ powersystems/nep2012/generation_12.pdf (18 June 2013): 47.

Table 3.7 Parameters to be adopted for 660 and 800 MW supercritical units

Parameter	660 MW	800 MW
Main steam pressure (kg/cm^2)	247	247
Main steam temperature (°C)	538/565	565
Reheat temperature (°C)	565/593	593

Source CEA 2012: CEA, "National Electricity Plan: Generation", January 2012, at: http://www.cea.nic.in/reports/powersystems/nep2012/generation_12.pdf (18 June 2013): 47, at: http://www.cea.nic.in/reports/powersystems/nep2012/generation_12.pdf (18 June 2013) MoF
°C = degree Celsius; cm = centimetre; kg = kilogramme; MW = megawatt

3.4.2.1 Supercritical Technology

With rapidly expanding thermal generation capacity, the installation of large-sized supercritical units is being encouraged in India to enhance the efficiency of coal-fired thermal generation, and thereby reduce coal consumption and GHG emissions and achieve faster capacity additions. An efficiency gain of about 2 percent over that of subcritical units is achieved with the adoption of higher temperature and pressure parameters,[37] thus reducing carbon emissions. While some supercritical units of 660 MW have already been commissioned during the 11th FYP, it is estimated that nearly 50 percent of total coal-based thermal capacity addition during the 12th FYP (2012–2017) would be through supercritical units.[38] In the 13th FYP (2017–2022) meanwhile, coal-based thermal capacity addition under the UMPP scheme will be only through supercritical units.[39] Further details of the parameters being adopted in India for supercritical units are given below (Table 3.7).

As a number of supercritical units are planned in the coming years, the Indian government has made efforts to foster the indigenous manufacturing of equipment and components for supercritical plants by attracting international manufacturers to establish such production facilities in India. This is also expected to bring down the costs associated with the setting up of such plants. For instance, a proposal for the bulk ordering of 11 supercritical units of 660 MW has been approved by the government for various projects of the National Thermal Power Corporation (NTPC) and Damodar Valley Corporation (DVC).[40] The proposal envisages mandatory conditions of the suppliers setting up manufacturing of supercritical units in the country, thus ensuring the initial few orders that would be vital for new joint ventures being set up to kick start indigenous manufacturing.

Several joint ventures and collaborations between Indian companies and international manufacturers such as L&T-MHI Boilers Private Limited (between

---

[37] Ibid: 47.

[38] Ibid: 46–47.

[39] Ibid: 47.

[40] "Bulk Ordering of 660 MW Supercritical Units: Award of Supercritical Units Deferred till January 2011" (1 August 2010), at: http://prosperingindianpowersector.blogspot.in/2010/08/bulk-ordering-of-660-mw-supercritical.html (13 July 2013).

Larsen & Toubro Limited (L&T), India, and Mitsubishi Heavy Industries Limited [MHI]), Toshiba JSW Turbine and Generator Pvt. Ltd. (between Toshiba Corporation [Toshiba] and JSW Group [JSW]), Alstom Bharat Forge Power Ltd. (ABFPL; between Alstom Power Holdings SA and Bharat Forge Ltd.), Thermax Babcock & Wilcox Energy Solutions Private Limited (TBWES; between Thermax Limited, India, and Babcock & Wilcox Power Generation Group, Inc.), and ACB India (between Ansaldo Caldaie and Gammon India) are now involved in the manufacture of supercritical power equipment in the country.[41]

3.4.2.2 Ultra Supercritical Technology

While there is no standard definition for ultra supercritical power plants, plants adopting higher steam temperatures of 600/600 °C are defined as one.[42] Experience of this technology is limited to a few countries such as Japan, Germany and Denmark. Ultra supercritical technology offers additional efficiency gain of about 0.75 percent over 800 MW supercritical units.[43] Apart from techno-economics and suitability for Indian coal, mitigation of the effects of GHG emissions and reducing the cost of power generation are other major factors providing added impetus to the adoption of this technology. Among notable partnerships in the country in this field is the collaboration between Bharat Heavy Electricals Ltd. (BHEL), a leading power plant manufacturer, and Indira Gandhi Centre for Atomic Research (IGCAR) to develop ultra supercritical steam generators.[44]

Table 3.8 shows the reduction achieved in CO_2 emissions with supercritical and ultra supercritical parameters in coal-based thermal power generation.

3.4.2.3 Integrated Gasification Combined Cycle Technology

IGCC is widely regarded as one of the leading clean coal technologies. Presently, the efficiency of IGCC plants is comparable to that of supercritical pulverised coal plants. For large-sized plants, IGCC technology can potentially achieve higher efficiencies than pulverised coal technology, primarily due to the higher efficiency of combined cycle systems. For this very same reason, IGCC plants also have lower CO_2 emissions. Although India is pursuing its own research for the development of IGCC technology suitable for Indian coal, the technology's efficiency

---

[41] CEA, "National Electricity Plan: Generation", January 2012, at: www.cea.nic.in/reports/powersystems/nep2012/generation_12.pdf (18 June 2013): 48.

[42] Ibid.

[43] Ibid: 46.

[44] Ibid: 47–48.

[45] CEA, "National Electricity Plan: Generation", January 2012, at: www.cea.nic.in/reports/powersystems/nep2012/generation_12.pdf (18 June 2013): 48.

Table 3.8 Comparison of the performance parameters of subcritical, supercritical and ultra supercritical thermal power plants

Performance	Subcritical	Supercritical	Ultra supercritical
Net heat rate (kcal/kWh)	2,380	2,233	1,986
Net efficiency (HHV; %)	36.1	38.5	43.3
Coal use (MT/year)	1.469	1.378	1.221
CO_2 emitted (MT/year)	3.30	3.09	2.74
CO_2 emitted (g/kWh)	894	830	738

Source Bindra 2010: Bindra, G.S., "Super Critical Technology and Use of Higher Thermal Unit Ratings For Future Capacity Addition Programme", Paper presented at 4th Thermal Power India 2010, New Delhi, India, 28–29 January 2010, at: http://www.indiacoreevents.in/bulletin/papers-tpi2010/G-S-Bindra-Bhel-Supercritical-Technology-And-Use-Of-Higher-Thermal-Unit-Ratings.pdf (13 July 2013): 39, at: http://www.indiacoreevents.in/ bulletin/papers-tpi2010/G-S-Bindra-Bhel-Supercritical-Technology-And-Use-Of-Higher- Thermal-Unit-Ratings.pdf (13 July 2013)

CO_2 = carbon dioxide; HHV = higher heating value; g = gramme; kcal = kilocalories; kWh = kilowatt hour; MT = million tonnes

Note Assumptions—500 MW net plant output; 85 percent capacity factor; pulverised coal combustion technologies without carbon capture and storage (CCS)

under Indian ambient conditions is expected to be lower than that achieved with low-ash international coal. Among ongoing IGCC projects in India is a 125 MW IGCC demonstration plant, which is being developed at Vijayawada by BHEL and Andhra Pradesh Power Generation Corporation (APGENCO).[45]

3.4.2.4 Ultra Mega Power Projects

In order to augment the country's power generating capacity and improve the efficiency of power generation, the GoI in association with the CEA has launched an initiative for the development of nine coal-based UMPPs—each with a capacity of about 4,000 MW—through a competitive bidding process. The UMPPs will use supercritical technology, thus helping to reduce GHG emissions as well as minimise costs due to the economies of scale. As added incentive, the GoI has extended full exemption from central excise and custom duties for goods procured for setting up these UMPPs.

---

[46] CEA, "Hydro Development Plan for 12th Five Year Plan (2012–2017)", at: http://cea.nic.in/reports/hydro/hydro_develop_12th_plan.pdf (19 July 2013): 2.

[47] CEA, "Monthly Review of Power Sector Performance", March 2012, at: http://www.cea.nic.in/reports/monthly/inst_capacity/mar12.pdf (13 July 2013): 15.

3.4.3 Developing Conventional Sources

3.4.3.1 Accelerated Development of Hydro Power

Assessment studies by the CEA have identified hydroelectric potential of 84,400 MW at 60 percent plant load factor, corresponding to 150,000 MW of installed capacity.[46] Till 31 March 2012, however, only 38,990 MW potential had been harnessed.[47] In order to meet the growing demand for power in the country without affecting CO_2 emissions, it is essential that the remaining hydro potential be fully harnessed too. With this in view, a 50,000 MW hydroelectric initiative was launched by the Prime Minister of India.[48] As part of the initiative, 162 schemes of approximately 50,000 MW capacity were identified at one-go to facilitate the faster development of hydropower.[49] The Integrated Energy Policy[50] of the GoI envisages the development of the country's entire hydro potential by year 2031–2032.[51] All efforts are afoot to rapidly realise India's full hydro potential.

3.4.3.2 Nuclear Power

The Indian nuclear power programme is a viable alternative to fossil fuel-fired power stations and a possible answer to India's growing energy demand in the long term. Its three-stage nuclear power development programme[52] envisages the utilisation of India's indigenous resources through a fuel-linked sequential. The first phase, which consists of conventional nuclear reactors, involves pressurised heavy water reactors (PHWRs) using natural uranium. As of 31 March 2011, 18 PHWRs, totalling 4,460 MW capacity, were in operation in the country and the construction of four PHWRs of 700 MW capacity each was in progress.[53]

The second phase consists of fast breeder reactors (FBRs) using plutonium-based fuel that is obtained by reprocessing spent fuel from first-stage reactors. The

---

[48] "PM Launches 50,000 MW Hydro Power Initiative", in: *The Economic Times* (24 May 2003), at: http://articles.economictimes.indiatimes.com/2003-05-24/news/27537712_1_hydro-power-power-sector-mw-additional-capacity (19 July 2013).

[49] Ibid.

[50] Planning Commission, "Integrated Energy Policy: Report of the Expert Committee", August 2006, at: http://planningcommission.gov.in/reports/genrep/rep_intengy.pdf (29 July 2013).

[51] Ibid: 41.

[52] NPCIL, "Nuclear Power: India's Sustainable Route to Low-carbon Energy", Twenty-fourth Annual Report 2010–11, at: http://large.stanford.edu/courses/2012/ph241/bordia1/docs/annual_report2010_2011.pdf (19 July 2013): 48.

[53] Ibid: 64.

[54] Ibid: 48.

[55] Ibid.

first FBR has been launched with the construction of a 500 MW prototype FBR (PFBR) at Kalpakkam, Tamil Nadu.[54] A successful transition to the third phase, which will use thorium-based reactors, will require a reprocessing of the spent fuel from the second stage. For the third phase, which is presently in the pilot stage, development of technology is being pursued.[55]

In India, Nuclear Power Corporation of India Limited (NPCIL), which is a public sector enterprise wholly owned by the GoI and under the administrative control of the Department of Atomic Energy (DAE), undertakes the design, construction, operation and maintenance of nuclear power stations for the generation of electricity in pursuance of the schemes and programmes of the GoI under provisions of the Atomic Energy Act, 1962.[56]

At present, India has 4,780 MW capacities in operation, which includes eighteen PHWRs totalling 4,460 MW capacity. Two light water reactors (LWRs) of 1,000 MW capacity each and one FBR of 500 MW capacity are under construction.[57] India has plans to indigenously construct 10 PHWRs of 700 MW each and two FBRs of 500 MW each.[58] It also plans to add 40,000 MW capacity in the form of LWRs with international cooperation by the year 2032.[59]

Post-Fukushima Daiichi: India's Position

The catastrophe at the Fukushima I Nuclear Power Plant at Daiichi in the aftermath of the Richter 9.0 earthquake on 11 March 2011 and the subsequent tsunami with waves over 14 m high[60] has forced a rethink of the future of nuclear power in the energy mix of countries. The Atomic Energy Regulatory Board (AERB) of India has in response constituted special task forces to study the implications of the accident at Fukushima to ascertain the need for enhancing the safety of nuclear plants in the country. A detailed walk down of all the plants in India has been conducted and their preparedness to deal with such situations is being ensured. Recommendations of the AERB are being implemented for safety. Additionally, the World Association of Nuclear Operators (WANO) has issued guidelines in the form of Significant Operating Experience Reports (SOERs).

---

[56] Ibid: 5.

[57] Ibid: 64.

[58] Ibid: 64–65.

[59] Ibid: 84.

[60] Ibid: 39.

[61] For the location of operating nuclear power plants in India and those being constructed or considered, see ibid: 12. For more information on the seismic zone mapping of India, see "India Seismic Zone Map", at: http://www.mapsofindia.com/maps/india/seismiczone.htm (29 July 2013).

[62] NPCIL, "Chairman's Statement: 26th Annual General Body Meeting-2013", 5 July 2013, at: http://npcil.nic.in/pdf/CMD_Statement_2013.pdf (29 July 2013): 5.

While evaluating the suitability and safety of nuclear power for India, it is important to bear a few facts in mind. Unlike in Japan, the seismic situation in India is different—most existing and proposed sites for nuclear power plants in India lie in areas of moderate seismic activity.[61] In keeping with the regulatory requirement that specifies the absence of active seismic faults in the vicinity of a nuclear installation, India has a large coastline with moderate seismic activity. Also, instead of private corporations, India's nuclear facilities are all run by its government. Not only this, the reactors currently being constructed and those under consideration for the future will rely heavily on passive features for critical core cooling functions that will circumvent the need for human intervention.[62] It is expected that the construction of safer reactors with passive core cooling features will inspire more public confidence and spur the acceptance of nuclear power stations.

In the face of its impending energy crisis and the threat of climate change, it is imperative that India continues to pursue its nuclear power options while developing and deploying inherently safer state-of-the-art plants. The addition of nuclear power plants will reduce India's dependence on coal-based plants, and thereby reduce its GHG emissions. All indications are that India will pursue such a course.

3.4.3.3 Gas Turbine-Based Power

Since gas turbines use cleaner fuels such as natural gas, liquefied natural gas (LNG), distillate oil and naphtha, gas turbine-based plants are a more environment friendly mode of power generation.[63] With this in view, the setting up of several new gas-based combined cycle power plants has been proposed.[64] However, in spite of its many advantages, the limited availability of gas is a major issue that has adversely affected the operation of existing plants and the development of new gas-based stations in India. Currently, LNG is being imported into the country to meet demand.

---

[63] CEA, "National Electricity Plan: Generation", January 2012, at: www.cea.nic.in/reports/powersystems/nep2012/generation_12.pdf (18 June 2013): 101, 107.

[64] Ibid: 107.

[65] CEA, "Monthly Review of Power Sector Performance", December 2012, at: http://www.cea.nic.in/reports/monthly/inst_capacity/dec12.pdf (13 July 2013): 15.

[66] MNRE, "Wind: Wind Power Programme", at: http://www.mnre.gov.in/schemes/grid-connected/solar-thermal-2/ (28 July 2013).

Table 3.9 Tentative programme for grid-interactive renewable power during the 12th Five-Year Plan

Source/system	Target (MW)
Wind power	11,000
Biomass power, baggasse cogeneration and biomass gasifiers	2,100
Small hydro (up to 25 MW)	1,600
Solar power	3,800
Total	18,500

Source MoP (2012: 29)
MW = megawatt

3.4.4 Harnessing Non-conventional Sources: Renewable Energy

Renewable energy can potentially play a significant role in India, and all efforts are being taken by the GoI to harness this potential. India's installed capacity from renewable energy sources as of 31 December 2012 was 25,856 MW.[65] The Ministry of New and Renewable Sources of Energy (MNRE) has proposed capacity addition for grid-interactive renewable power from sources such as wind, biomass and small hydro during the 12th FYP (Table 3.9).

India's current potential for power generation from wind is estimated at 48,500 MW.[66] Among the various renewable technologies, wind technology in India has matured following the enactment of the Act. India's installed wind turbine generating capacity as of June 2013 stands at 19,565 MW.[67] This growth is a result of government support—while the central government has provided incentives in the form of capital and interest subsidies, accelerated depreciation, tax holidays, concessional custom duty on imports and generation-based incentives,[68] state governments have encouraged wind generation in the country by allotting land at concessional rates, and providing sales and electricity duty/tax exemption/reduction.[69]

---

[65] CEA, "Monthly Review of Power Sector Performance", December 2012, at: http://www.cea.nic.in/reports/monthly/inst_capacity/dec12.pdf (13 July 2013): 15.

[66] MNRE, "Wind: Wind Power Programme", at: http://www.mnre.gov.in/schemes/grid-connected/solar-thermal-2/ (28 July 2013).

[67] MNRE, "Achievements", at: http://www.mnre.gov.in/mission-and-vision-2/achievements/ (29 July 2013).

[68] MNRE, "Frequently Asked Questions on Wind Power Programme (FAQ)", at: http://mnre.gov.in/file-manager/UserFiles/faq_wind.pdf (29 July 2013).

[69] Ibid.

3.4.5 Enhancement of the Power Infrastructure

3.4.5.1 Lowering Transmission and Distribution Losses

The average aggregate technical and commercial (AT&C) losses for utilities selling directly to consumers at the national level, which were 26.58 percent in the year 2009–2010, have marginally decreased to 26.15 percent in the year 2010–2011.[70] These losses are proposed to be brought down to a level of 15 percent in 5 years with the help of Restructured Accelerated Power Development & Reforms Programme (R-APDRP) by various power utilities.[71] This lowering would amount to a reduction of T&D losses of about 10 percent on the all-India level. Many of the distribution networks of state utilities are old, weak and overloaded. These networks need to be strengthened and revamped for improved quality and reliability of supply and for reduction in losses. Reducing such losses would automatically imply that correspondingly less capacity addition is needed and a reduction in GHG emissions is effected.

3.4.5.2 Renovation and Modernisation of Old Power Stations

The renovation and modernisation (R&M) of thermal power stations and the life extension (LE) of an existing old power station improves efficiency, thus resulting in reduced GHG emissions, additional generation at lower cost in the short term, and improved availability, safety and reliability.

There is immense potential for improving the operating efficiency of coal-fired thermal power stations by way of R&M. India's approach towards R&M schemes and plans has undergone a paradigm shift lately, and the focus is now more on performance optimisation and generation maximisation. However, earlier attempts at efficiency improvement and the reduction of auxiliary power consumption in the country did not receive the attention and priority they deserved due to a variety of reasons.[72]

The CEA has prepared a National Enhanced Efficiency Renovation and Modernization Program for implementation during the 11th and 12th FYPs. During the 11th FYP, the programme envisages R&M of 18,965 MW capacity and LE of 7,318 MW while R&M of 4,971 MW and LE of 16,532 MW is targeted during the 12th FYP.[73]

---

[70] PFC, "Report on 'The Performance of State Power Utilities for the Years 2008-09 to 2010-11'", at: http://www.pfcindia.com/writereaddata/userfiles/file/ResearchReport/Performance_Report_State_Power_Utilities_forfy_2008-09to2010-11_03102012.pdf (20 July 2013): 4-iii.

[71] Ibid: 15.

[72] CEA, "National Electricity Plan: Generation", January 2012, at: www.cea.nic.in/reports/powersystems/nep2012/generation_12.pdf (18 June 2013): 49–50.

[73] Ibid: 51.

3.4.5.3 Phasing Out of Old and Small-Sized Generating Units

A very large number of small-sized units of 100 MW capacity or less are in operation in the country.[74] While these units, which are of the non-reheat type, have very low design efficiencies inherently, many of them are performing at efficiencies much lower than accounted for by their design due to age and technological obsolescence.[75] The average plant load factor of these units is very low, and such units will need to be phased out over the next decade or so. During the 11th FYP, 3,000 MW capacity from coal/lignite-based units of unit size lesser than 100 MW was targeted to be retired.[76] Aiming for further GHG emissions reduction, the gradual winding down of more inefficient units is planned during the 12th and 13th FYPs.[77]

3.4.6 Energy Conservation and Energy Efficiency

A national movement for energy conservation could significantly reduce the need for fresh investment into energy supply systems in the coming years. It is imperative that all-out efforts are made at all levels to realise this potential. Energy conservation is a cause to which all citizens can contribute. Be it a household or factory, a small shop or large commercial building, a farmer or office worker, every user and producer of energy can and must make this effort for their own benefit as well as that of the nation's.

Efficient use of energy is the primary objective in the present context and the GoI has already enacted the Energy Conservation Act, 2001 (EC Act), which provides much needed legal framework and institutional arrangement for the promotion of energy efficiency.[78] Under the EC Act, thermal power stations along with other major energy-intensive industries have already been declared as 'designated consumers'.[79] The following are some of the relevant provisions under the EC Act for power stations that have been designated consumers:

- to designate or appoint an energy manager in charge of the activities for energy efficiency and conservation;
- to have an accredited energy auditor conduct energy audits;

---

[74] MoP, "Guidelines for Renovation and Modernisation/Life Extension Works of Coal/Lignite Based Thermal Power Stations", October 2009, at: http://cea.nic.in/reports/renov_modern/th_randm_annex-a.pdf (16 July 2013): 5.

[75] Ibid.

[76] CEA, "National Electricity Plan: Generation", January 2012, at: www.cea.nic.in/reports/powersystems/nep2012/generation_12.pdf (18 June 2013): 53.

[77] Ibid: 53–54.

[78] MoP, "The Energy Conservation Act, 2001", at: http://powermin.nic.in/acts_notification/energy_conservation_act/ (29 July 2013).

[79] Ibid.

- to furnish information with regard to energy consumed and the actions taken on the recommendations of accredited energy auditors; and,
- to comply with the norms of energy consumption.

3.5 National Action Plan on Climate Change

India's National Action Plan on Climate Change (NAPCC),[80] released by the Prime Minister of India on 30 June 2008, recognises the need to maintain a high growth rate for increasing the living standards of the vast majority of its people and reducing their vulnerability to the impacts of climate change. The Indian Prime Minister Dr. Manmohan Singh while launching the NAPCC, described the seriousness of India's climate change challenges as:

Our vision is to make India's economic development energy-efficient. Over a period of time, we must pioneer a graduated shift from economic activity based on fossil fuels to one based on non-fossil fuels and from reliance on non-renewable and depleting sources of energy to renewable sources of energy. In this strategy, the sun occupies centre-stage, as it should, being literally the original source of all energy. We will pool our scientific, technical and managerial talents, with sufficient financial resources, to develop solar energy as a source of abundant energy to power our economy and to transform the lives of our people. Our success in this endeavour will change the face of India. It would also enable India to help change the destinies of people around the world.[81]

The NAPCC outlines eight national missions, representing multipronged, long-term and integrated strategies for achieving certain key goals in the context of climate change.[82] These missions are:

- The Jawaharlal Nehru National Solar Mission (JNNSM);
- National Mission for Enhanced Energy Efficiency (NMEEE);
- National Mission on Sustainable Habitat;
- National Water Mission;
- National Mission for Sustaining Himalayan Ecosystem (NMSHE);
- National Mission for a "Green India";
- National Mission for Sustainable Agriculture (NMSA); and,
- National Mission on Strategic Knowledge for Climate Change (NMSKCC).

---

[80] Prime Minister's Council on Climate Change, "National Action Plan on Climate Change", at: http://pmindia.nic.in/climate_change_english.pdf (20 June 2013).

[81] MNRE, "Jawaharlal Nehru National Solar Mission: Towards Building 'Solar India'", at: http://www.mnre.gov.in/file-manager/UserFiles/mission_document_JNNSM.pdf (21 July 2013): 6.

[82] Prime Minister's Council on Climate Change, "National Action Plan on Climate Change", at: http://pmindia.nic.in/climate_change_english.pdf (20 June 2013): 2–5.

In order to achieve a sustainable development path that simultaneously advances economic and environmental objectives, the NAPCC is guided by the following principles[83]:

- protecting the poor and vulnerable sections of society through an inclusive and sustainable development strategy, sensitive to climate change;
- achieving national growth objectives through a qualitative change in direction that enhances ecological sustainability, leading to further mitigation of GHG emissions;
- devising efficient and cost-effective strategies for end-user DSM;
- deploying appropriate technologies for both adaptation and mitigation of GHG emissions extensively as well as at an accelerated pace; and,
- engineering new and innovative forms of market, regulatory and voluntary mechanisms.

3.5.1 National Solar Mission

India being a tropical country, sunshine is available for longer hours each day and at great intensity. Solar energy therefore has great potential as a future energy source for the country. Solar energy also has the advantage of permitting the decentralised distribution of energy, thereby empowering people at the grassroots level.

To address India's energy security and to meet the global challenge of climate change, the GoI has launched a special solar mission, known as JNNSM or the National Solar Mission under the brand name 'Solar India', to establish India as a global leader in solar energy by creating the policy conditions for its diffusion across the country as quickly as possible.[84] The mission has set a target of 20,000 MW and stipulates the implementation of this target in three phases by 2022 (Phase I: 2012–2013; Phase II: 2013–2017; Phase III: 2017–2022 [or the 13th FYP]).[85] In order to facilitate the acceptability of grid-connected solar power generation, the first phase will involve the bundling of relatively expensive solar power with power generated from coal-based stations.

To achieve its various targets, the mission aims to:

- create an enabling policy framework for the deployment of 20,000 MW of solar power by 2022;
- ramp up capacity of grid-connected solar power generation to 1,000 MW within 3 years by 2013 and an additional 3,000 MW by 2017 through the mandatory

---

[83] Ibid: 2.

[84] CEA, "National Electricity Plan: Generation", January 2012, at: www.cea.nic.in/reports/powersystems/nep2012/generation_12.pdf (18 June 2013): 62.

[85] Ibid.

use of renewable purchase obligations by utilities backed by preferential tariffs. This capacity can be more than doubled, reaching 10,000 MW of installed power by 2017 or more based on enhanced and enabled international finance and technology transfer. The target for 2022, which is 20,000 MW or more, is dependent on the 'learning' of the first two phases, which if successful, could lead to conditions of grid-competitive solar power. This transition could be appropriately scaled up, based on the availability of international finance and technology;

- create favourable conditions for solar manufacturing capability, particularly solar thermal for indigenous production and market leadership;
- promote programmes for off-grid applications, reaching 1,000 MW by 2017 and 2,000 MW by 2022;
- achieve a solar thermal collector area of 15 million sq. metres by 2017 and 20 million sq. metres by 2022; and,
- deploy 20 million solar lighting systems for rural areas by 2022.

3.5.2 National Mission for Enhanced Energy Efficiency

The NMEEE is an India-government initiative proposed to address national problems of inefficient energy use.[86] Market-based approaches are proposed to unlock energy efficiency opportunities, which have been estimated to be around US$13.5 billion.[87] India's various schemes aim to achieve annual fuel savings in excess of 23 mtoe, which would lead to CO_2 emission mitigation of 98 million tons per year.[88]

Para 4.2 of the NAPCC mandates[89]:

- Perform Achieve and Trade (PAT)—A market-based mechanism to enhance the cost effectiveness of improvements in energy efficiency in energy-intensive large industries and facilities through the certification of energy savings that can be traded;
- Market Transformation for Energy Efficiency (MTEE)—Accelerating the shift to energy efficient appliances in designated sectors through innovative measures to make products more affordable;
- Energy Efficiency Financing Platform (EEFP)—Creation of mechanisms that would help finance DSM programmes in all sectors by capturing future energy savings;

---

[86] CEA, "National Electricity Plan: Generation", January 2012, at: www.cea.nic.in/reports/powersystems/nep2012/generation_12.pdf (18 June 2013): 213–215.

[87] Garnaik, S.P., "National Mission for Enhanced Energy Efficiency", at: http://www.moef.nic.in/downloads/others/Mission-SAPCC-NMEEE.pdf (28 July 2013): 4.

[88] Ibid.

[89] Prime Minister's Council on Climate Change, "National Action Plan on Climate Change", at: http://pmindia.nic.in/climate_change_english.pdf (20 June 2013): 3.

- Framework for Energy Efficient Economic Development (FEEED)–Developing fiscal instruments to promote energy efficiency;
- under the PAT scheme, eight industrial sectors, namely power, iron and steel, fertiliser, cement, aluminium, pulp and paper, textile and chlor-alkali, have been included. These eight industries will be required to achieve a reduction of Specific Energy Consumption (SEC) from their baseline SECs within 3 years (i.e., during the period from 2011–2012 to 2013–2014). This is expected to save nearly 10 mtoe by 2013–2014, out of which around 67 percent will be accounted for by the power sector.[90]

3.5.3 National Mission for a "Green India"

Forests contribute much to the preservation of ecological balance and biodiversity of a country. Accordingly, the National Mission for a "Green India" or the Green India Mission (GIM) was launched to enhance India's ecosystems, including its carbon sinks. The afforestation of six million hectares of forest/non-forest land is planned under the campaign, which will result in 50–60 million tonnes of enhanced annual sequestration of CO_2 by the year 2020, at a total cost of around US$9 billion for the entire mission over the next decade.[91]

3.5.4 National Clean Energy Fund

Harnessing renewable energy sources to reduce dependence on fossil fuels is well recognised as a sound strategy for combating global warming and climate change. The National Clean Energy Fund (NCEF) has been established to provide a positive push to the development of clean energy for funding research and innovative projects in clean energy technologies. In order to finance the fund, a clean energy cess on coal produced or imported in India has been introduced, which levies Rs. 50 (or approximately US$1) per tonne of coal.[92] This cess, which is expected to yield approximately Rs. 27,500 million (or approximately US$444 million) annually, would be used for financing and promoting clean energy initiatives and funding research.[93]

---

[90] Garnaik, S.P., "National Mission for Enhanced Energy Efficiency", at: http://www.moef.nic.in/downloads/others/Mission-SAPCC-NMEEE.pdf (28 July 2013): 6.

[91] "National Mission for a Green India", at: http://www.naeb.nic.in/documents/GIM_Brochure_26March.pdf (28 July 2013): 2.

[92] MoF, "Subject: Levy of Clean Energy Cess – Regarding", F.No.354/72/2010-TRU, 24 June 2010, at: http://www.cbec.gov.in/excise/cx-circulars/cx-circulars-10/circ-cec01-2k10.htm (16 July 2013).

[93] "Clean Coal Cess Makes Power Costlier", in: *mjunction* (20 July 2010), at: http://mjunction.in/market_news/coal_1/clean_coal_cess_makes_power_co.php (16 July 2013).

3.6 Conclusion

India faces formidable challenges for maintaining and enhancing its economic growth, which is vital for lifting the lives of millions of its people living in dire poverty, and combating the threat of climate change. India has taken the climate change challenge very seriously and is making all possible efforts to reduce GHG emissions, with several policy initiatives being taken to mitigate the effects of emissions. The Prime Minister's Council on Climate Change has decided that India will work simultaneously on several fronts in a focussed manner to deal with the many challenges of climate change. The launching of missions such as JNNSM, NMEEE and GIM, among others, are important steps to mitigate GHG emissions at the national level. Enhancing energy efficiency, energy conservation, harnessing hydro potential and renewable sources of energy, deploying state-of-the-art technologies to reduce emission levels during power generation, and establishing nuclear-based generation are some of the efforts India is making to ensure its energy security while mitigating climate change.

References

CEA (Central Electricity Authority), 2008: *All India Electricity Statistics (2006–07): General Review, 2008* (New Delhi: Central Electricity Authority).

CEA (Central Electricity Authority), 2010: *Report on Eighteenth Electric Power Survey of India* (New Delhi: Central Electricity Authority).

CEA (Central Electricity Authority), 2012: *All India Electricity Statistics: General Review, 2012* (New Delhi: Central Electricity Authority).

IEA (International Energy Agency), 2010: *World Energy Outlook 2010* (Paris: OECD/IEA).

MIB (Ministry of Information and Broadcasting), 2013: "Land and the People", in: *India 2013* (New Delhi: Publication Division, Ministry of Information and Broadcasting).

MoP (Ministry of Power), "Demand for Power and Generation Planning", in: *Report of the Working Group on Power for Twelfth Plan (2012–17)*, January 2012 (New Delhi: Ministry of Power), at: http://planningcommission.nic.in/aboutus/committee/wrkgrp12/wg_power1904.pdf (18 July 2013).

WCED (World Commission on Environment and Development), 1987: *Our Common Future* (Oxford: Oxford University Press).

Abbreviations

ABFPL	Alstom Bharat Forge Power Ltd
Act	The Electricity Act, 2003
AERB	Atomic Energy Regulatory Board
AIMA	All India Management Association
APGENCO	Andhra Pradesh Power Generation Corporation
APTEL	Appellate Tribunal for Electricity
AT&C	Aggregate technical and commercial

AWG-LCA	Ad Hoc Working Group on Long-term Cooperative Action under the Convention
BHEL	Bharat Heavy Electricals Ltd
CCS	Carbon capture and storage
CDM	Clean Development Mechanism
CEA	Central Electricity Authority
CFBC	Circulating fluidised bed combustion
CO_2	Carbon dioxide
DAE	Department of Atomic Energy
DNA	Designated National Authority
DSM	Demand-side management
DVC	Damodar Valley Corporation
EC Act	The Energy Conservation Act, 2001
EEFP	Energy Efficiency Financing Platform
EIA U.S.	Energy Information Administration
EPS	Electric Power Survey
FBR	Fast breeder reactor
FEEED	Framework for Energy Efficient Economic Development
FYP	Five-Year Plan
GCV	Gross calorific value
GDP	Gross domestic product
GHG	Greenhouse gas
GIM	Green India Mission
GoI	Government of India
GW	Gigawatt
GWh	Gigawatt hour
IEA	International Energy Agency
IEEE	Institute of Electrical and Electronics Engineers
IEI	Institution of Engineers (India)
IET	Institution of Engineering and Technology
IEX	Indian Energy Exchange
IGCAR	Indira Gandhi Centre for Atomic Research
IGCC	Integrated Gasification Combined Cycle
INAE	Indian National Academy of Engineering
JNNSM	Jawaharlal Nehru National Solar Mission
JSW	JSW Group
Kg	Kilogramme
kWh	Kilowatt hour
KWU	Siemens Power Generation Group
L&T	Larsen & Toubro Limited
LE	Life extension
LNG	Liquefied natural gas
LWR	Light water reactor
MHI	Mitsubishi Heavy Industries Limited

MIB	Ministry of Information and Broadcasting, India
MNRE	Ministry of New and Renewable Sources of Energy
MoEF	Ministry of Environment and Forests, India
MoF	Ministry of Finance, India
MoP	Ministry of Power, India
MTEE	Market Transformation for Energy Efficiency
mtoe	Million tonnes of oil equivalent
MW	Megawatt
NAMA	Nationally Appropriate Mitigation Action
NAPCC	National Action Plan on Climate Change
NCDMA	National CDM Authority
NCEF	National Clean Energy Fund
NEP	National Electricity Policy
NMEEE	National Mission for Enhanced Energy Efficiency
NMSA	National Mission for Sustainable Agriculture
NMSHE	National Mission for Sustaining Himalayan Ecosystem
NMSKCC	National Mission on Strategic Knowledge for Climate Change
NPCIL	Nuclear Power Corporation of India Limited
NTPC	National Thermal Power Corporation
OECD	Organisation for Economic Co-operation and Development
PAT	Perform Achieve and Trade
PFBR	Prototype fast breeder reactor
PFC	Power Finance Corporation Ltd.
PHWR	Pressurised heavy water reactor
PTC	PTC India Ltd. (formerly Power Trading Corporation of India Ltd.)
R&D	Research and development
R&M	Renovation and modernisation
R-APDRP	Restructured Accelerated Power Development and Reforms Programme
RGGVY	Rajiv Gandhi Grameen Vidyutikaran Yojana
SEC	Specific Energy Consumption
SERC	State Electricity Regulatory Commission
SOER	Significant Operating Experience Report
T&D	Transmission and distribution
Toshiba	Toshiba Corporation
UMPP	Ultra Mega Power Projects
UNFCCC	United Nations Framework Convention on Climate Change
WANO	World Association of Nuclear Operators
WCED	World Commission on Environment and Development
WEC	World Energy Council

Chapter 4
Climate Change and Energy Security Post-Cancún: The Indonesia Perspective

Fitrian Ardiansyah, Neil Gunningham and Peter Drahos

Abstract Indonesia faces a huge challenge in continuing the development of its economy while reducing its greenhouse gas (GHG) emissions at the same time. Using examples from the energy sector, which is the backbone of the country's economy as well as one of the largest contributors of GHG emissions, this chapter examines the climate-energy nexus in Indonesia, taking into account the various relevant policies, political contexts and institutions that have influenced the climate change and energy discourses in the country. A particular focus will be given to efforts that have tried to balance actions on climate change mitigation while addressing energy security. The chapter critically reviews policy efforts to integrate climate change objectives into energy policy, and discusses policy impediments and implementation gaps that prevent the country from smoothly integrating its climate change and energy security objectives, with the development of geothermal energy in Indonesia taken as a case study.

F. Ardiansyah (✉)
WEH Stanner Room #1.38, Crawford School of Public Policy, The Australian National University (ANU), 132 Lennox Cross, Canberra ACT 0200, Australia
e-mail: fitrian.ardiansyah@anu.edu.au
URL: http://fitrianardiansyah.com

N. Gunningham
Climate and Environmental Governance Network, National Research Centre for Occupational Health and Safety (OHS) Regulation, RegNet, Research School of Pacific and Asian Studies, and Fenner School of Environment and Society, The Australian National University (ANU), Canberra ACT 0200, Australia
e-mail: neil.gunningham@anu.edu.au
URL: http://www.anu.edu.au/fellows/ngunningham

P. Drahos
Centre for Governance of Knowledge and Development, RegNet, Research School of Pacific and Asian Studies, The Australian National University (ANU), Canberra ACT 0200, Australia
e-mail: peter.drahos@anu.edu.au
URL: http://www.anu.edu.au/fellows/pdrahos

N. A. Putra and E. Han (eds.), *Governments' Responses to Climate Change: Selected Examples From Asia Pacific* 10, SpringerBriefs in Environment, Security, Development and Peace, DOI: 10.1007/978-981-4451-12-3_4, © The Author(s) 2014

Keywords Climate change · Energy governance · Energy policy · Geothermal energy · Renewable energy

4.1 Introduction

Indonesia is perceived as a nation that has made a strong pledge to tackle climate change. This has been increasingly the case since 2007, when the country hosted the 13th session of the Conference of the Parties (COP 13) to the United Nations Framework Convention on Climate Change (UNFCCC) that historically marked the Bali Action Plan (BAP) and the beginning of 2 years of formal negotiations until the Copenhagen Climate Change Conference (COP 15) in 2009 (Purnomo et al. 2013: 1–2). One famous milestone was the commitment made by the Indonesian President Susilo Bambang Yudhoyono at a G20 meeting on 25 September 2009, stating that his government was devising a policy to cut emissions by 26 percent by the year 2020 from 'business as usual' (BAU) levels, and up to 41 percent with international support (Melisa 2010: 1). Beyond Cancún COP 16 (and until the latest COP 18 in Doha), Indonesia appears to have been maintaining the public position it took in Bali by issuing domestic policies and programmes and establishing institutions aimed at helping the country to achieve its climate change targets.

Indonesia's policies and programmes have been a mix of initiatives that aim to address emissions from deforestation and changes in land use, including from forest and land fires, as well as step up investment in the energy sector, especially to boost energy efficiency and renewable energy. In 2011, the Presidential Regulation No. 61 of 2011 on the National Action Plan for Greenhouse Gas Emissions Reduction (Rencana Aksi Nasional Penurunan Emisi Gas Rumah Kaca (RAN-GRK))[1] was enacted to translate the President's commitment and serve as an umbrella policy for climate change mitigation actions. This policy is expected to further help mainstream climate change-related initiatives into Indonesia's development agenda.

In specific sectors, for instance, to address deforestation and peatland conversion, Indonesia has issued the Presidential Instruction No. 10 of 2011 on Suspension of Granting New Licenses and Improvement of Natural Primary Forest and Peatland Governance (commonly known as the moratorium on primary forest

---

[1] Kementerian PPN/BAPPENAS, "The Presidential Regulation of the Republic of Indonesia No. 61 Year 2011 on the National Action Plan for Greenhouse Gas Emissions Reduction" (October 2011), at: http://www.bappenas.go.id/get-file-server/node/11521/ and http://www.bappenas.go.id/get-file-server/node/11522/ (9 January 2013).

and peatland conversion)[2] as part of its commitment to Norway.[3] With regard to
the climate-energy nexus, policy works and outputs have been available at least
since the mid-2000s. These include the Presidential Regulation No. 5 of 2006 on
National Energy Policy,[4] which stipulates an energy mix for 2025 that lowers the
country's dependence on oil and significantly increases the role of new and
renewable energy, and Law No. 30 of 2007 on Energy,[5] which highlightes the
increasing need for new and renewable energy as well as energy conservation.

Institutions have also been set up to foster the implementation of these policies,
including the creation of the National Council on Climate Change (Dewan Nasional Perubahan Iklim [DNPI])[6] in 2008, the National Energy Council (Dewan
Energi Nasional (DEN))[7] in 2008, the Indonesia Climate Change Trust Fund
(ICCTF),[8] the REDD+ (Reducing Emissions from Deforestation and Forest

---

[2] Kementerian Dalam Negeri (The Ministry of Home Affairs), "Instruksi Presiden Republik
Indonesia No. 10 Tahun 2011 tentang Penundaan Pemberian Izin Baru dan Penyempurnaan Tata
Kelola Hutan Alam Primer dan Lahan Gambut" (Instruction by The President of the Republic of
Indonesia No. 10 Year 2011 regarding Suspension of Granting New Licenses and Improvement
of Natural Primary Forest and Peatland Governance) (19 September 2011), at: http://
www.depdagri.go.id/media/documents/2011/09/19/i/n/inpres_no.10-2011.pdf and for the Eng-
lish translation, at: http://www.daemeter.org/wp-content/files/INPRES-10_2011__EN.pdf (9
January 2013).

[3] In May 2010, the governments of Indonesia and Norway signed a letter of intent (LoI) regarding
cooperation on reducing greenhouse gas emissions from deforestation and forest degradation. Based
on this LoI, the Indonesian government agrees to issue a policy on forest and peatland conversion
moratorium. See CIFOR, "Letter of Intent between the Government of the Kingdom of Norway and
the Government of the Republic of Indonesia on 'Cooperation on Reducing Greenhouse Gas
Emissions from Deforestation and Forest Degradation'" (May 2010), at: http://
www.forestsclimatechange.org/fileadmin/photos/Norway-Indonesia-LoI.pdf (9 January 2013).

[4] MEMR, "Peraturan Presiden Republik Indonesia No. 5 Tahun 2006 tentang Kebijakan Energi
Nasional" (The Presidential Regulation of the Republic of Indonesia No. 5 of 2006 on National
Energy Policy) (2008), at: http://prokum.esdm.go.id/perpres/2006/perpres_05_2006.pdf and for
the English translation, at: http://faolex.fao.org/docs/pdf/ins64284.pdf (9 January 2013).

[5] MEMR, "Undang-Undang Republik Indonesia No. 30 Tahun 2007 tentang Energi" (Law of
the Republic of Indonesia No. 30 Year 2007 on Energy) (2008), at: http://prokum.esdm.go.id/uu/
2007/uu-30-2007-en.pdf (9 January 2013).

[6] Dewan Nasional Perubahan Iklim (DNPI) was established by the President with the Presidential
Regulation No. 46 of 2008. See MEMR, "Peraturan Presiden Republik Indonesia No. 46 Tahun
2008 tentang Dewan Nasional Perubahan Iklim" (The Presidential Regulation of the Republic of
Indonesia No. 46 of 2008 on National Council on Climate Change) (in Bahasa Indonesia) (2008),
at: http://prokum.esdm.go.id/perpres/2008/Perpres-46-2008.pdf (9 January 2013).

[7] Dewan Energi Nasional (DEN) was established by the President with the Presidential Regulation
No. 26 of 2008. See MEMR, "Peraturan Presiden Republik Indonesia No. 26 Tahun 2008 tentang
Pembentukan Dewan Energi Nasional" (The Presidential Regulation of the Republic of Indonesia
No. 26 of 2008 on The Establishment of National Energy Council) (in Bahasa Indonesia) (2008), at:
http://prokum.esdm.go.id/perpres/2008/Perpres-26-2008.pdf (9 January 2013).

[8] The Indonesia Climate Change Trust Fund (ICCTF) was established by BAPPENAS in 2009.
See Kementerian PPN/BAPPENAS, "Persiapan pembentukan Trustee ICCTF" (The Preparation
of the Establishment of the ICCTF Trustee) (October 2009), at: http://bappenas.go.id/node/116/
2417/persiapan-pembentukan-trustee-icctf—(9 January 2013).

Degradation-Plus) Task Force,[9] the Secretariat of RAN-GRK,[10] and the Directorate General of New, Renewable Energy, and Energy Conservation (DG-EB-TKE).[11] Such policies and institutions, endeavouring to translate the President's pledges into action, appear ambitious. This may reflect the President's desire to improve Indonesia's international profile as a constructive player on the global stage, particularly in light of its membership of the G20 (Jotzo 2012: 94). Indonesia appears to be one of those countries that would like to show that developing countries can also contribute to climate change mitigation (Ardiansyah 2010: 56–58).

Whatever the motivation, these commitments have (perhaps intentionally) generated momentum for climate change mitigation at a national level. It is, however, one thing to make ambitious pledges and another to implement them. As Jotzo argues (2012: 94), pledges do not necessarily equal action in the context of Indonesia. The gap between the Indonesian government's climate change aspirations and the situation on the ground as well as the outcomes that have been achieved thus far may be a large one. Objections and protests are likely to occur against a particular policy that may furthermore hinder its implementation. In a democratic country such as Indonesia, substantive reform may face resistance from various interest groups (Jotzo 2012: 94). Democratic countries with large populations that include many poor people cannot move on policy in the way that authoritarian regimes can.[12] In the energy sector, for example, to mitigate greenhouse gas (GHG) emissions from fossil fuel combustion, Indonesia seems to be struggling, particularly when it comes to eliminating fuel and electricity subsidies,[13] which up until now have fostered inefficient use of fuel and electricity in the country (Nurdianto/Resosudarmo 2011: 117).

It would therefore be interesting to comprehend whether it is technically, economically and politically possible for Indonesia's economic development to continue even as its emissions are being reduced. This chapter,[14] using examples from the energy sector (which is not only the backbone of Indonesia's economy

---

[9] The REDD+ (Reducing Emissions from Deforestation and Forest Degradation-Plus) Task Force was established by the President Decree No. 19 of 2010.

[10] The Secretariat of Rencana Aksi Nasional Penurunan Emisi Gas Rumah Kaca (RAN-GRK) was established to facilitate the implementation of RAN-GRK as stipulated in the Presidential Regulation No. 61 of 2011.

[11] The Directorate General of New, Renewable Energy, and Energy Conservation (DG-EBTKE) was established as part of the mandate of the Presidential Regulation No. 24 of 2010.

[12] In contrast to democratic states, as analysed by Desai et al. (2009: 106), political liberalisation is negatively related in authoritarian states.

[13] Ardiansyah, Fitrian, "Bearing the Consequences of Indonesia's Fuel Subsidy", in: *East Asia Forum* (4 May 2012), at: http://www.eastasiaforum.org/2012/05/04/26135/ (9 January 2013).

[14] This chapter benefits greatly from a three-year study of energy governance and its implications for climate change in Indonesia conducted by Gunningham and Drahos, the second and third authors. The purpose of the study was to understand how policy and regulatory frameworks established at the national level actually steer energy policy, and how energy policy is coordinated internationally.

but also one of its largest contributors of GHG emissions), examines the climate-energy nexus in the country, taking into account the relevant policies, political contexts and institutions that have influenced its climate change and energy discourses. A particular focus will be given to efforts that have tried to balance actions on climate change mitigation while addressing energy security. Section two of the chapter looks at Indonesia's energy profile and policy, among others, providing the latest update on policy efforts to integrate climate change objectives into its energy policy, such as policies that push for the phasing out of fossil fuels and the acceleration of renewable energy development. In section three, the chapter discusses further the policy impediments and implementation gaps that prevent the country from smoothly integrating its climate change and energy security objectives, with the development of geothermal energy taken as a case study. In this section, the chapter examines some structural obstacles and highlights initiatives that might serve as good solutions. The final section discusses further steps and concludes the chapter.

4.2   Energy: Options and Policies Vis-à-Vis Climate Change Objectives

Energy plays an important role in boosting a country's economy and social welfare. It supplies the fuels that transform raw materials in production systems for domestic and export commodities. This process has created many economic benefits for Indonesia. For instance, oil- and gas-related revenues accounted for as much as 27 percent of total government revenues in 2004–2009[15] and contributed US $34.4 billion to state revenue in 2011 (PwC 2012: 5). Although Indonesia has been considered a net oil importer since 2004,[16] it is still expecting an increase in oil and gas investments.[17]

With the growth of Indonesia's economy, its energy demands will increase.[18] According to the Ministry of Finance (2009: 5), total energy demand is growing by

---

[15] DBS Group Research, "Economics Indonesia: 2011 Budget Preview" (1 November 2010), at: https://www.dbsvresearch.com/research%5Cdb%5Cresearch.nsf/(vwAllDocs)/ 67A56A7116FF77AB482577CE003053AD/%24FILE/id_2010Nov1.pdf (16 March 2011).

[16] In 2004, oil production averaged 1,100 thousand barrels per day (tbd) while consumption hit 1,200 tbd (Sa'ad 2009: 4391–4392).

[17] In 2011, more than 30 new oil and gas contracts were entered into in Indonesia (PwC 2012: 5). In 2013, according to the Energy and Mineral Resources Minister, Indonesia expects an increase in oil and gas investments by as many as 274 work plans, with a value of US $26.2 billion. See, Azwar, Amahl S., "Indonesia Expects an Increase in Oil and Gas Investment for 2013", in: *The Jakarta Post* (8 January 2013), at: http://www.thejakartapost.com/news/2013/01/08/indonesia-expects-increase-oil-and-gas-investment-2013.html (9 January 2013).

[18] While some scholars believe that there is a causal relationship between energy consumption and economic growth (Asafu-Adjaye 2000), others argue that there is no such causality between the two (Jafari et al. 2012).

around 7 percent per year, as the transport and industrial sectors grow, and as households become more affluent. Another report shows that energy consumption has steadily and rapidly increased in Indonesia since 1990 at the rate of 3.5 percent per year.[19]

In Indonesia, this annual significant demand for energy has generally been met using fossil fuels. Too much dependency on fossil fuels, however, has created its very own set of problems for the country. For instance, the Vice Minister of the Ministry of Energy and Mineral Resources (MEMR) Rubi Rubiandini stated that high energy consumption has caused and could accelerate the imbalance between the exploitation of fossil energy resources (such as oil, gas and coal) and the speed of inventing new reserves, leading to a depletion of Indonesia's reserves and increasing dependency on imported energy.[20] This is particularly the case with oil, where the high global oil prices and the dependency on imported oil have placed considerable strain on the Indonesian economy, which is further exacerbated by its heavy oil and electricity subsidies. According to Indonesia's Director General of Treasury, Ministry of Finance, Agus Suprijanto, the realisation of energy subsidies cost the government Rp. 306.5 trillion (US $31.5 billion) in 2012, a figure much higher than that projected in the latest government's adjusted budget (Rp. 202.35 trillion or US $20.8 billion).[21] As shown in Fig. 4.1, this figure is a significant increase from the cost incurred in 2010 (Rp. 139.9 trillion or US $14.4 billion) and that projected in the original budget of 2012.[22]

Despite oil being depleted in Indonesia, the country remains extremely resource rich and a major player in the world energy economy, particularly due to gas and coal. In 2011, Indonesia overtook Australia as the world's largest coal exporter (although Australia remains the world's largest supplier of coking coal), with its exports amounting to over 300 million tonnes (mt).[23] This situation may change in the near future as Indonesia's domestic demand for coal is likely to increase strongly, affecting its exports, as suggested by the Indonesian Energy and Minerals

---

[19] With the exception of 1997, when Indonesia was hit by the Asian financial crisis, primary energy consumption has been increasing steadily and rapidly. See ABB, "Indonesia Energy Efficiency Report" (10 January 2011), at: http://www05.abb.com/global/scot/scot316.nsf/veritydisplay/1a65dd16a3c538acc125786400514251/%24file/indonesia.pdf (9 January 2013).

[20] MEMR, "25% of Indonesian Have No Access to Energy" (21 December 2012), at: http://www.esdm.go.id/news-archives/general/49-general/6123-25-of-indonesian-have-no-access-to-energy.html (9 January 2013).

[21] Noviani, Ana, "Subsidi Energi: Realisasi Belanja Membengkak Dari Pagu APBN-P 2012" (Energy Subsidy: A Significant Increase from the Government Adjusted Budget), in: *Bisnis* (3 January 2012), at: http://www.bisnis.com/articles/subsidi-energi-realisasi-belanja-membengkak-dari-pagu-apbn-p-2012 (9 January 2013).

[22] The original budget (before adjusted) for energy subsidies in 2012 was Rp168.5 trillion (US $17.3 billion), or approximately 17 % of the government's total original budget. See Ministry of Finance, "Budget Statistics 2006–2012", in: *Indonesian Energy Electricity Sheet* (2012), at: http://energy-indonesia.com/08data/BudgetStatistics2006-2012.pdf (1 December 2012).

[23] WCA, "Coal Market and Transportation" (2012), at: http://www.worldcoal.org/coal/market-amp-transportation/ (9 January 2013).

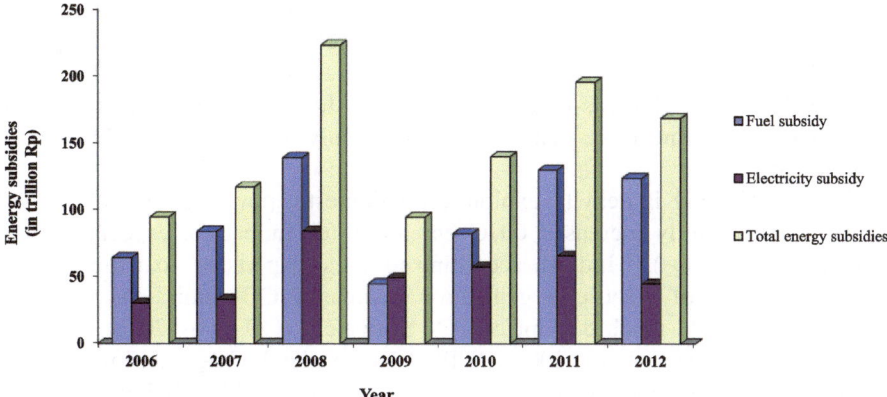

Fig. 4.1 Energy (fuel and electricity) subsidies in the central government budget (2006–2012). *Source* Adapted from Ministry of Finance 2012. The original budget (before adjusted) for energy subsidies in 2012 was Rp168.5 trillion (US $17.3 billion), or approximately 17 percent of the government's total original budget. See Ministry of Finance, "Budget Statistics 2006–2012", in: *Indonesian Energy Electricity Sheet* (2012), at: http://energy-indonesia.com/08data/ BudgetStatistics2006-2012.pdf (1 December 2012): 4, 11, at: http://energy-indonesia.com/08 data/BudgetStatistics2006-2012.pdf

Minister Jero Wacik.[24] In particular, coal is the fuel of choice for boosting electricity generation. This was evidenced by the government's 'crash programme', as mandated by the Presidential Regulation No. 71 of 2006,[25] instructing the state-owned electricity company (Perusahaan Listrik Negara [PLN]) to accelerate the development of 10,000 megawatts (MW) of coal-fired generating plants.

With regard to gas, as of January 2012, the country was the world's seventh largest exporter of natural gas[26] and the fourteenth largest holder of proven natural gas reserves.[27] However, as has occurred with coal, natural gas shortages caused

[24] Tanquintic-Misa, Esther, "Indonesia to Impose Duties on Coal Exports, Face up to $11B in Export Earning Losses", in: *International Business Times* (6 June 2012), at: http://au.ibtimes.com/ articles/348998/20120606/indonesia-coal-exports.htm#.UO6bmm_Zbzk (9 January 2013).

[25] "Peraturan Presiden Republik Indonesia No. 71 Tahun 2006 tentang Penugasan kepada PT Perusahaan Listrik Negara [PLN] untuk Melakukan Percepatan Pembangunan Pembangkit Tenaga Listrik yang Menggunakan Batubara" (Regulation of the President of the Republic of Indonesia Number 71 Year 2006 Concerning Assignment to Pt. Perusahaan Listrik Negara (Persero) to Accelerate the Development of Electric Power Generation Using Coal), at: http:// www.presidenri.go.id/DokumenUU.php/239.pdf (9 January 2013) and for the English translation, at: http://repit.files.wordpress.com/2011/11/perpres-71-tahun-2006-english-version.pdf (25 June 2013).

[26] "Thematic Map: Natural Gas—Exports—World", in: *Index Mundi* (1 January 2012), at: http://www.indexmundi.com/map/?v=138 (9 January 2013).

[27] EIA, "Indonesia Energy Profile: Reorienting Away from Exports to Serve Domestic Consumption—Analysis", in: *EurasiaReview* (10 January 2013), at: http:// www.eurasiareview.com/10012013-indonesia-energy-profile-reorienting-away-from-exports-to-serve-domestic-consumption-analysis/#.UO6bqW_Zbzk (10 January 2013).

by production problems and rising domestic consumption could further reduce the resource available for export in Indonesia.[28] To accelerate the use of gas for power generation, the President issued the Presidential Regulation No. 4 of 2010[29] that instructs PLN to develop gas-fired power stations, along with power plants that use renewable energy and more coal (known commonly as the 'second-phase crash programme').

However, if coal is heavily exploited for domestic power generation, it will result in significantly increased GHG emissions for Indonesia, especially from carbon dioxide (CO_2).[30] Indeed, according to some experts, if coal remains the dominant source of electricity generation, increased CO_2 emissions from the power sector would reach 810 mt of CO_2 equivalent (CO_2e) by 2030, or nearly seven times the levels seen in 2005 (DNPI 2010: 25). High CO_2 emissions are also likely to exacerbate global climate change impacts to which Indonesia, as an archipelagic nation, is already vulnerable (Ministry of Environment 2010: 14–15). Not only this, the projected growth in emissions resulting from coal-fired plants would be in contradiction with the President's commitments to reduce GHG emissions, as legally stipulated in the Presidential Regulation No. 61 of 2011.

Against the backdrop of increasing global oil prices and depleted reserves, and in conjunction with the projected increase in GHG emissions from coal-fired plants, other energy sources—and, in particular, renewable energy and gas—do appear increasingly attractive from Indonesia's perspective, both for reasons of energy security as well as climate change mitigation. Natural gas, for example, can serve as a 'short-term bridge' fuel that can act as a substitute for not only oil-based power plants but also ageing coal-fired power plants.[31] In the second-phase crash programme,[32] for instance, the Indonesian government is planning to build

---

[28] Ibid.

[29] Presiden Republik Indonesia, "Peraturan Presiden Republik Indonesia No. 4 Tahun 2010 tentang Penugasan kepada PT Perusahaan Listrik Negara [PLN] untuk Melakukan Percepatan Pembangunan Pembangkit Tenaga Listrik yang Menggunakan Energi Terbarukan, Batubara dan GAs" (The Presidential Regulation of the Republic of Indonesia No. 4 of 2010 on The Tasks Given to PLN to Accelerate the Development of Power Plants based on Renewable Energy, Coals and Gas) (2011), at: http://www.presidenri.go.id/DokumenUU.php/408.pdf (9 January 2013).

[30] For further discussion on coal and its negative environmental impacts, see Ardiansyah et al. (2012: 97–98).

[31] Kirkland, Joel; Climatewire, "Natural Gas Could Serve as 'Bridge' Fuel to Low-carbon Future", in: *Scientific American* (25 June 2010), at: http://www.scientificamerican.com/article.cfm?id=natural-gas-could-serve-as-bridge-fuel-to-low-carbon-future&page=3 (9 January 2013).

[32] The second-phase crash programme, to be realised during the period 2009–2018, was introduced to push for the development of coal-fired power plants, gas-fired power plants and geothermal power plants at a projected cost of US $21.3 billion. See Mansur, S., "Comprehensive Study on: Crash Program Progress and National Electricity Business Opportunity 2008–2015 (Featuring Exploration & Exploitation Renewable Energy Sources)", in: *Media Data Riset* (March 2008), at: http://www.scribd.com/doc/19295066/CRASH-PROGRAM-PROGRESS-AND-NATIONAL-ELECTRICITYpdf (10 December 2010).

1,660 MW of gas combined-cycle plants.[33] Likewise, although for a different purpose, Indonesia was able to successfully carry out a programme that facilitated a shift from kerosene to liquid petroleum gas as the principal fuel for cooking (Pertamina and WLPGA 2012: 10–18). Between 2007 and mid-2012, this programme provided 53.9 million conversion packages to Indonesian citizens, covering nearly 54 million households and small and medium enterprises (SMEs), or 93 percent of its target (Pertamina/WLPGA 2012: 14–15).

It should be noted, however, that while a 'bridge fuel', gas is still a fossil fuel albeit with GHG emissions approximately half that of coal.[34] What is more, the prolonged use of coal and gas could undermine serious efforts that aim to shift the country's energy mix to more renewables.[35] A crucial factor in this context is the changing nature of the world gas market. Technologies to extract gas and oil from shale rock will allow North America to become a major exporter of both, with the IEA (2012: 23) predicting that the United States will by around 2020 become the world's biggest oil producer. One implication of such technological revolution in fossil fuel extraction is that countries such as Indonesia would benefit from much cheaper world gas prices. While a long-term decline in gas and coal prices may provide Indonesia with a measure of energy security,[36] for the world as whole this development would come at the cost of climate insecurity. Not surprisingly, the implication of changes in global gas prices for Indonesia remains the subject of much research.

In principle, renewable energy has considerable attractions. Indonesia has large potential to harness geothermal energy in addition to moderate prospects of generating hydro and, to a lesser extent, solar power. Nonetheless, there are many obstacles to this scenario, including the prolonged use of coal and gas that would be necessary before Indonesia would be able to harness renewables and achieve a low-carbon future.

For Indonesia, it is clearly a monumental challenge to provide reliable, affordable and sustainable energy while contributing to climate change mitigation. Poverty remains widespread, with only 72.95 percent of Indonesian people having

---

[33] Alfian, "Indonesia to Construct 93 Power Plants in 2nd Project", in: *The Jakarta Post* (30 January 2010), at: http://www.thejakartapost.com/news/2010/01/30/indonesia-construct-93-power-plants-2nd-project.html (9 January 2013).

[34] The percentage of GHG emissions from natural gas would be higher if one accounts for the impact of leakage (including and especially of methane) during the extraction process, and particularly from the extraction of shale gas. See Howarth et al. (2011: 660, 683–685).

[35] The International Energy Agency (IEA) has argued that the tripling of output by 2035 from unconventional gas sources such as shale gas could end support for renewables. See Harvey, Fiona, "'Golden Age of Gas' Threatens Renewable Energy, IEA Warns", in: *The Guardian* (29 May 2012), at: http://www.guardian.co.uk/environment/2012/may/29/gas-boom-renewables-agency-warns (9 January 2013).

[36] According to OECD/IEA (2012: 11), production of unconventional gas, primarily shale gas, is predicted to more than triple to 1.6 trillion cubic metres in 2035. Large increases are expected to occur in the United States, Russia, China, Australia, India, Canada and Indonesia (OECD/IEA 2012: 11).

access to electricity in 2011.[37] The mitigation of poverty remains therefore a high domestic priority at the national level and realising this objective may conflict with the country's plans for renewable energy, at least in the short term, especially if renewable energy development does not take into account the poverty equation. Renewable energy, in general, could definitely play a bigger role in helping to address energy security and providing access to electricity. Some studies positively show that renewable energy sources could meet up to 35 percent[38] of Indonesia's energy needs by 2035 (Leitmann et al. 2009: 62–78; Marpaung et al. 2012: 5–6). However, with the current energy mix still being dominated by fossil fuels,[39] it is likely to be difficult for renewable energy sources to compete, especially at a larger scale. Coal in particular, is cheap, readily accessed domestically, and readily converted into electricity by a proven and relatively quickly installed technology in the form of coal-fired power stations. The Vice Minister of MEMR reflected on this matter by saying that, although there is big potential for utilising renewable energy, the current use of such energy is relatively small and fossil fuel energy utilisation still dominates national energy consumption.[40] Switching to and speeding up the development of renewable energy may require substantive financial support and further policy reforms.

Of the various renewable energy sources (for example hydro, solar, biomass and geothermal) available in Indonesia, geothermal power, in particular, appears to be the most appropriate and preferred solution. This is understandable given that the country holds approximately 40 percent of the world's geothermal reserves,

---

[37] Winoto, Ashary Teguh; Marliska, Elif Doka; Prasetyo, M. Himawan; Simangunsong, Sahat, "Rural Electrification in Indonesia: Target and Development", Country Report 2012, in: *Energy Indonesia*, at: http://energy-indonesia.com/03dge/05chiho.pdf (9 January 2013).

[38] At the 2nd Congress of the East Asian Association of Environmental and Resource Economics (EAAERE), in Bandung, Indonesia, Marpaung et al. (2012: 1, 5–6) presented an Asian-Pacific Integrated Model (AIM)/End-use model developed to examine the energy security implications of a renewable portfolio standard (RPS) in Indonesia. In their model, three levels of RPS are considered—15, 25 and 35 percent—within the planning horizon 2005–2035.

[39] As of 2010, Indonesia's energy mix consisted of 49.7 percent oil, 20.1 percent gas, 24.5 percent coal and 5.7 percent renewables. See "Indonesia's Energy Security and Geothermal Development", in: *Indonesia Soken* (July 2012), at: http://www.indonesiasoken.com/pdf/FREE_046_20120730_energypercent20report.pdf (9 January 2013). According to the Presidential Regulation No. 5 of 2006, new and renewable energy's share is expected to be 17 percent by 2025. The MEMR has expanded this government target and set a 25 percent share of new and renewable energy in the 2025 energy mix, known as 'Energy Vision 25/25'. See Azahari, Hasrul Laksamana, "New and Renewable Energy Policies", in: *Energy Indonesia* (18 July 2012), at: http://energy-indonesia.com/03dge/03.pdf (9 January 2013).

[40] MEMR, "25 percent of Indonesian Have No Access to Energy" (21 December 2012), at: http://www.esdm.go.id/news-archives/general/49-general/6123-25-of-indonesian-have-no-access-to-energy.html (9 January 2013).

which are currently underutilised.[41] According to the Head of Geological Agency at MEMR Sukhyar (2011: 11), Indonesia's total potential geothermal resources and reserves are estimated at 28,994 megawatts-electrical (MWe), with an installed capacity of 1,196 MWe (accounting for approximately 4 percent of its total resources and reserves).

Should geothermal energy be exploited properly, it could compete[42] with fossil fuels in the power sector, helping Indonesia to meet its energy security and climate change mitigation objectives. Geothermal energy, as argued by Mackay (2008: 98–99, 234), has the potential to replace coal-fired power plants as a baseload electricity source with virtually no emissions. Considering such potential, the government appears to be serious in pushing for this form of renewable energy. The government issued Law No. 27 on Geothermal in 2003,[43] followed by the Government Regulation No. 59 of 2007 on Geothermal Business Activities[44] and the MEMR Regulation No. 14 of 2008[45] on Basic Tariff for the Selling of Electricity from Geothermal Power (PwC 2011: 23). Darma et al. (2010: 1–2) argue that the aforementioned policies, together with the MEMR Regulation No. 5 of 2009 on Energy Price,[46] have provided a strong incentive for geothermal energy development in the country and for achieving the government targets of 6,000 MW by 2020 and 9,500 MW by 2025.

This initiative was assisted by a newer tariff that was eventually introduced in January 2011 by the PLN, which involved an 18 percent tariff hike ceiling as a temporary measure to resolve tariff discrepancies and was in line with the MEMR Regulation No. 7 of 2010.[47] Since 2011, Indonesia's feed-in regulation has

---

[41] However, the President has announced plans to become the world's leading geothermal nation, with 44 plants to be built by 2014 with a capacity of 4,000 MW and capacity rising to 9,000 MW by 2025. See Allard, Tom, "Indonesia Goes to Ground for Energy", in: *Sydney Morning Herald* (1 May 2010), at: http://www.smh.com.au/environment/energy-smart/indonesia-goes-to-ground-for-energy-20100430-tzbv.html (16 March 2011).

[42] Given the effect of shale gas on gas prices as well as the world and Indonesia, the role and contribution of renewable energy in Indonesia may change. This remains the subject of extensive research.

[43] MEMR, "Undang-Undang Republik Indonesia No. 27 Tahun 2003 tentang Panas Bumi" (Law of the Republic of Indonesia No. 27 Year 2003 Concerning Geothermal), at: http://prokum.esdm.go.id/uu/2003/uu-27-2003.pdf (9 January 2013) and for the English translation, at: http://repit.files.wordpress.com/2011/09/uu-no-27-2003-english-version.pdf (25 June 2013).

[44] See FAOLEX, "Indonesia: Government Regulation No. 59/2007 Concerning Geothermal Business Activity" (5 November 2007), at: http://faolex.fao.org/cgi-bin/faolex.exe?rec_id= 070870%26database=faolex%26search_type=link%26table=result%26lang=eng%26format_ name=%40ERALL (15 July 2013).

[45] This regulation was subsequently replaced by MEMR Regulation No. 5 of 2009.

[46] MEMR, "Peraturan Menteri Energi Dan Sumber Daya Mineral Nomor: 05 Tahun 2009" (in Bahasa Indonesia), at: http://www.djlpe.esdm.go.id/modules/_website/files/37/File/permen-esdm-05-2009(1).pdf (15 July 2013).

[47] MEMR, "Peraturan Menteri Energi Dan Sumber Daya Mineral Nomor 07 Tahun 2010" (in Bahasa Indonesia), at: http://prokum.esdm.go.id/permen/2010/Permen%20ESDM%2007% 202010.pdf (15 July 2013).

provided a significantly higher tariff for geothermal power at US $9.7 cents per kilowatt hour (cents/kWh), although the scope was limited to geothermal fields included in the second-phase crash programme. In August 2012, the government increased the tariff again, based on the MEMR Regulation No. 22 of 2012.[48],[49] The Energy and Mineral Resources Minister believed that such strong support to renewable energy development was necessary to offset the entrenched position of oil in the country's electricity generation.[50] The Director General of New, Renewable Energy and Energy Conservation at MEMR Kardaya Warnika further argued that such feed-in-tariffs (FITs) was one of the policies that would provide price assurance to investors.[51] With the newest tariff, the government seems to hope that companies will have incentives to invest in geothermal energy.

If the share of geothermal energy increases in the national energy mix, GHG emissions from the power sector could be reduced, if not avoided, as several studies have shown (Marpaung et al. 2012: 5–6; Wijaya/Limmeechokchai 2009: 7). A 2009 study, for instance, showed that an increase of 10 gigawatts (GW)—the total geothermal potential presently ready for commercial extraction, as reported by the World Bank (Leitmann et al. 2009: 64)—in geothermal energy capacity by 2025 would result in emission savings of approximately 58 mt of CO_2e (Wijaya/Limmeechokchai 2009: 7).[52] Energy scenarios developed by the MEMR exhibit relatively similar patterns (Ariati 2009: 8). According to the model developed by the MEMR, from 2011 onwards national emissions based on a geothermal scenario (where geothermal energy's capacity reaches 27 GW or 8.4 percent of the national energy mix) will decline significantly in comparison to emissions based on BAU levels (i.e. the high use of coal) (Ariati 2009: 8).

To accelerate the development of other renewable energy sources, similar policies have been introduced, aiming at increasing the tariffs for these renewables as well. For hydro power, the government increased the tariff to Rp. 656–1,004 (US $ 0.07–0.10) per kilowatt hour (kWh) while, for biomass, the government has

---

[48] MEMR, "Peraturan Menteri Energi Dan Sumber Daya Mineral Republik Indonesia Nomor: 22 Tahun 2012" (in Bahasa Indonesia), at: http://prokum.esdm.go.id/permen/2012/Permen%20ESDM%2022%202012.pdf (15 July 2013).

[49] Aiming to further encourage investment in geothermal energy, the government started obliging state power producer PLN to pay a tariff on electricity generated from geothermal facilities at US $0.11 to nearly US $0.20 per kilowatt hour (kWh) depending upon the region where the resource was located. See Gipe, Paul, "Indonesia Launches "Crash" Renewables Program: Boosts Geothermal FITs", in: *Renewable Energy* (20 July 2012), at: http://www.renewableenergyworld.com/rea/news/article/2012/07/indonesia-launches-crash-renewables-program-boosts-geothermal-fits (9 January 2013).

[50] Ibid.

[51] MEMR, "Geothermal Feed in Tariff will Immediately be Signed" (5 July 2012), at: http://www.esdm.go.id/index-en/83-energy/5828-geothermal-feed-in-tariff-will-immediately-be-signed.html (9 January 2013).

[52] These CO_2 emissions savings are calculated by comparing the estimated amount of CO_2 emissions coming from geothermal power plants to that from coal-fired power plants.

increased the tariff to Rp. 975–1,398 (US $0.10–0.14) per kWh.[53] Given the huge potential[54] of hydro (i.e. 75,670 MW for large-scale and 769.69 MW for smaller-to-micro projects) and biomass (i.e. 49,810 MW), these policies are expected to considerably boost the development of renewable energy in the country. At the time of this writing, FITs for solar photovoltaic and wind were being formulated.[55]

To strengthen the policies that promote renewable energy, the government has also attempted to further reform its fossil fuel subsidies. These efforts are aimed at phasing out its dependency on imported and heavily subsidised oil. Throughout the past decade, the government has pursued various strategies to lessen the country's reliance on fossil fuels, including partially lifting the country's subsidies to the state oil company PERTAMINA (Perusahaan Tambang dan Minyak Negara),[56] the heavily petroleum product-dependent electricity sector, and for the consumption of petroleum products such as gasoline, diesel and kerosene (Beaton/Lontoh 2010: 17). Special Staff to the President for Climate Change Agus Purnomo argued in 2009 that cutting fossil fuel subsidies would be key to bolstering the renewable energy sector's competitiveness.[57] The results, however, have been mixed—while some initiatives can be considered a success, many of these remain politically contentious. The latest attempt in 2012 to increase petrol prices was rejected by the Indonesian parliament.[58] What the government subsequently did was to introduce a fuel quota in the market, although the quota is likely to be exceeded in 2012 compared to the original plan.[59]

Another set of policies that endeavours to integrate climate change objectives in the energy sector involves energy efficiency and conservation. There is a modest energy conservation component in the second-phase crash programme. In 2009, the Government Regulation No. 70 on Energy Conservation[60] was introduced with the aim of reducing technical losses in electricity generation, transmission and distribution. Households, for example, have been targeted with moves to adopt Energy Efficient Labelling and Standardisation, modelled on the Australian and

---

[53] Azahari, Hasrul L., "Indonesia's Feed-in Tariff for Renewable Energy", in: *Energy Indonesia* (14 May 2012), at: http://energy-indonesia.com/03dge/Hasrulpercent20L.%20Azahari.pdf (9 January 2013).

[54] Ibid.

[55] Ibid.

[56] PERTAMINA is Indonesia's state-owned oil and gas company. For the company's full profile, see "Company Profile", at: http://www.pertamina.com/index.php/home/read/company_profile (12 August 2011).

[57] Ardiansyah, Fitrian, "Bearing the Consequences of Indonesia's Fuel Subsidy", in: *East Asia Forum* (4 May 2012), at: http://www.eastasiaforum.org/2012/05/04/26135/ (9 January 2013).

[58] Ibid.

[59] Azwar, Amahl S., "Regulator Works to Keep Fuel Quota on Track", in: *The Jakarta Post* (14 December 2012), at: http://www.thejakartapost.com/news/2012/12/14/regulator-works-keep-fuel-quota-track.html (9 January 2013).

[60] Presiden Republik Indonesia, "Peraturan Pemerintah Republik Indonesia Nomor 70 Tahun 2009" (in Bahasa Indonesia), at: http://www.djlpe.esdm.go.id/modules/_website/files/36/File/PP%2070%202009.pdf (15 July 2013).

New Zealand standards, for electrical appliances (Hilmawan/Said 2009: 11). The President himself has set up a new team called the National Team on Energy and Water Conservation to promote energy conservation (Hilmawan/Said 2009: 8). The initial focus will be on the training and appointment of energy managers for large consumers. However, despite such policies, these initiatives are of a very limited nature. One of the reasons for this, as argued by Hilmawan/Said (2009: 9), is that the economic incentives provided for energy efficiency are not strong enough to promote the initiatives.[61]

Various policies and programmes have been formulated and introduced, which have provided a good platform for renewable energy development and energy conservation/efficiency in Indonesia. In reality, however, the realisation of renewable energy development is a long way from being smooth. Various obstacles have somehow slowed down, if not halted, the acceleration of this development. The forthcoming section discusses the factors that have prevented Indonesia from speeding up the implementation of this development, particularly for renewable energy. Political and institutional contexts are examined, and various options that may serve to provide at least partial solutions highlighted.

4.3 Impediments and Implementation Gaps in Renewable Energy Development: Geothermal Energy as Case Study

As elaborated in the previous section, the renewable energy potential of Indonesia is significant, as are the economic rewards from its realisation. However, in spite of increasingly changing its policies, the government has been largely perceived as lethargic in harnessing these renewable resources, with its energy policy remaining coal-centric.[62] The government faces a dilemma. On the one hand, there is constant pressure to meet the country's energy shortfall, which is immediate and pressing. On the other hand, waiting for renewable energy development to mature would mean that it might not be able to keep up with Indonesia's rapid energy demand in the short term and would have to face the discontent of energy consumers. For understandable reasons then, relying heavily on coal as a source of electricity generation has been viewed as an immediate solution for electricity generating capacity, one that also minimises the costly import of oil for the country. In the longer term, there are plausible energy security and climate change mitigation arguments for Indonesia to embrace a broader energy mix in which

---

[61] Compare, for example, the role of Energy Services Companies (ESCOs) in OECD countries, which provide structures that enable the industry to invest in energy efficiency without capital expenditure and are underpinned by banks.

[62] Wilcox, Jeremy, "Indonesia's Energy Transit: Struggle to Realize Renewable Potential", in: *Renewable Energy* (14 September 2012), at: http://www.renewableenergyworld.com/rea/news/article/2012/09/indonesias-energy-transit (9 January 2013).

renewable energy would play a much larger role. Yet, with the domination of coal and its attendant interest groups, one wonders whether renewable energy would really be able to find a level playing field. Governments in Indonesia are faced with the prospect of costly short-term politics from opposing fossil fuel lobbies and depriving the public of subsidy benefits in exchange for long-term emissions gains, moves whose political benefits will no doubt be in the long term as well. Under such calculus, most politicians would proceed cautiously.

There is also the issue of the existing grid. Indonesia's electricity generation and grid capacities are outdated and insufficient, operating at an average capacity factor of 66 percent.[63] Blackouts are an increasingly common problem (Bruch et al. 2011: 8, 19). In 2010, Indonesia's total on-grid generation capacity was 29 GW or around 0.12 kilowatts (kW) per capita.[64] Coal has substituted oil as the main energy source and accounts for 40 percent of the total capacity, with only around 10 percent coming from renewable energy sources (2.9 GW), mainly from large-scale hydro and geothermal power.[65] To boost renewable energy development, there is no option for the government but to expand its electricity infrastructure (i.e. grid development). It is reported that Indonesia, in 2011, only spent about 1.7 percent of its gross domestic product on infrastructure, compared to Malaysia's 5.4 percent and Thailand's 3.6 percent.[66] To continue its impressive economic growth, Indonesia needs to spend much more on infrastructure.[67] However, as a Badan Perencanaan Pembangunan Nasional (BAPPENAS; also known as National Development Planning Agency) energy planner/economist has pointed out, the construction of energy infrastructure requires thorough and consistent planning over a relatively long period.[68] He argues that Indonesia currently does not have any good long-term plans for the development of large-scale, complex energy infrastructure[69] (notwithstanding World Bank support in the form of a loan of US $225 million to strengthen transmission systems[70]). While the two strategic crash programmes described earlier involve government attempts to increase generating

---

[63] Differ Group, "The Indonesian Electricity System—A Brief Overview" (6 February 2012), at: http://www.differgroup.com/Portals/53/images/Indonesia_overall_FINAL.pdf (9 January 2013).

[64] Ibid.

[65] Ibid.

[66] Lontoh, Sonita, "A Smart Grid Can Help Sustain Indonesia's Growth", in: *The Jakarta Globe* (24 December 2012), at: http://www.thejakartaglobe.com/biscolumns/a-smart-grid-can-help-sustain-indonesias-growth/563357 (9 January 2013).

[67] Ibid.

[68] Hanan Nugroho, "Energy and Climate Change Management in Indonesia", in: The Jakarta Post (6 December 2009), at: http://www.thejakartapost.com/news/2009/06/12/energy-and-climate-change-management-indonesia.html (16 March 2011).

[69] Ibid.

[70] World Bank Indonesia, "New Financing to Help Meet Growing Electricity Demands of Indonesian Economy" (8 July 2010), at: http://web.worldbank.org/WBSITE/EXTERNAL/COUNTRIES/EASTASIAPACIFICEXT/INDONESIAEXTN/0,,contentMDK:22641183%7EmenuPK:224605%7EpagePK:2865066%7EpiPK:2865079%7EtheSitePK:226309,00.html (16 March 2011).

capability, coal is seen as centrally important,[71] with geothermal and gas playing at best supporting roles in the second phase of these programmes.

To date, the overall renewable energy sector in Indonesia has not expanded at any substantial pace. A senior government officer of BAPPENAS, however, argued that "as of November 2008, we had over 5 MW of grid connected renewable commissioned with over 86 MW being built. While micro-hydro is still by far the largest contributor of renewable resources, the government is now actively exploring the potential to utilize wind energy" (Girianna 2009: 2). With regard to geothermal energy, achieving the official government target (i.e. the target set by the Indonesian government and stipulated in its policy or regulation) to develop this energy has proven a challenge, as elaborated in the following studies. In 2010, a review commissioned by the MEMR found that it would be difficult to meet the official government target of building 3,967 MW of geo-thermal capacity by 2014, and that the most the government could hope to deliver was 2,297 MW (Castlerock 2010: iv). A 2008 World Bank report predicted sim-ilarly low figures, suggesting that only 2,800 MW of geothermal energy capacity would be installed by 2020 (World Bank 2008: 2). Figure 4.2 shows that, based on current trends, Indonesia can only build around 1,700 MW of geothermal capacity by 2014, 2,750 MW by 2020 and only 4,000 MW by 2025.[72] These projected figures are much lower than the government's targets.

One of the reasons for the slow progress of geothermal energy development is probably the government's incapacity to invest in research and development—it has neither the money nor the accessible skills to do so. It is only over the last 4 years that Indonesia has set up specific focus groups for geothermal energy, such as the one established at the Institute of Technology at Bandung, West Java.[73] Structural impediments remain important and include, among others, fossil fuel subsidies (as elaborated in the previous section), the division of power between the national government and the provinces (or the power struggle that has emerged in the decentralised era of Indonesia) and the seeming inability to create a risk environment that will attract commercial investment in renewable energy.

Although there has been progress in addressing these structural problems (including the adjustment to the tariff system, as described earlier), the deployment of renewable energy technology has been sluggish. Even with the adjusted price, there are a few regulatory issues that may need to be addressed, including whether the new pricing would automatically apply to projects won under the previous

[71] Mansur, S., "Comprehensive Study on: Crash Program Progress and National Electricity Business Opportunity 2008-2015 (Featuring Exploration & Exploitation Renewable Energy Sources)", in: *Media Data Riset* (March 2008), at: http://www.scribd.com/doc/19295066/CRASH-PROGRAM-PROGRESS-AND-NATIONAL-ELECTRICITYpdf (10 December 2010).

[72] The government's target and projected installed capacities were calculated based on the list of power plants under development as part of the Indonesian government's second phase crash programme of 10,000 MW.

[73] Institut Teknologi Bandung, "Background", Magister Program in Geothermal Technology (2013), at: http://geothermal.itb.ac.id/background (9 January 2013).

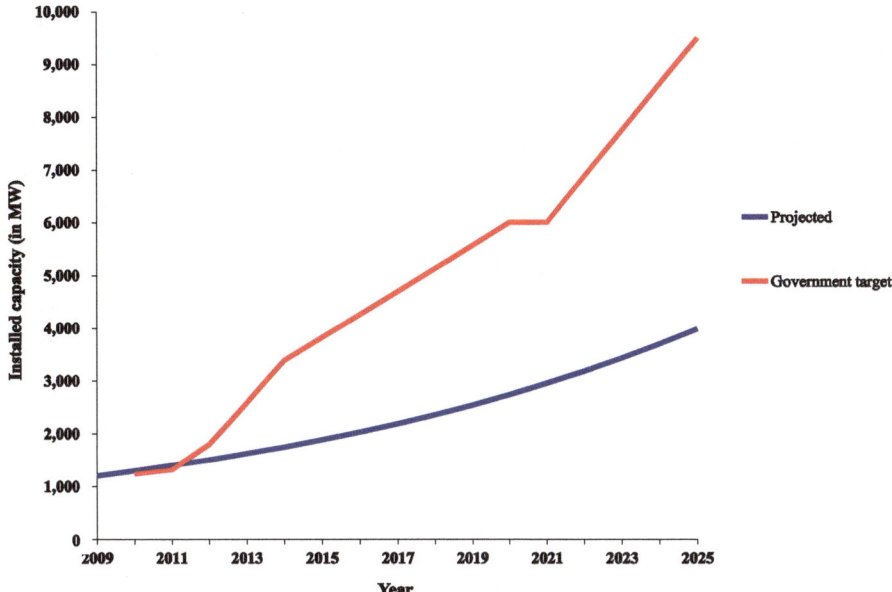

Fig. 4.2   Projection of geothermal installed capacity. *Source* Pozon et al. (2012): 95–107

auctioning scheme, which primarily awarded contracts to the lowest bidder. As a result of the previous scheme, contracts were awarded at prices that were far too low for development to go ahead. However, applying the new price retrospectively could spark a host of legal problems, including lawsuits brought forth by those who have lost out on previous auctions.[74]

Another immediate challenge is the fact that, since decentralisation following the collapse of Suharto's authoritative regime and as part of the push for demo-cratisation and regional autonomy, the decision-making process in Indonesia has become immeasurably more complex, as evidenced by the devolution of power that took place in 2001 to its 33 provinces and more than 400 districts. This devolution has important implications for energy policy as provincial and district governments have also been given the right and responsibility to issue concessions and operating licenses for renewable energy and energy efficiency. Most provinces and districts, however, have very limited capacity and only very limited understanding of the implications of various energy scenarios, and there is no established route or pro-cedure though which to pursue such initiatives[75] (Resosudarmo et al. 2010: 9–10).

---

[74] Indonesia—and, in this case, PERTAMINA and PLN—has faced litigation regarding geothermal energy development. In a famous case against Karaha Bodas company, the International Arbitration Institute in Switzerland ruled in favour of Karaha Bodas. See BIICL, "Case Note: Karaha Bodas and Himpurna Arbitrations" (2008), at: http://www.biicl.org/files/3931_2000_himpurna_and_karaha_bodas_arbitrations.pdf (9 January 2013).

[75] With the provincial governments issuing tenders for geothermal working areas, the central government lacks the proper authority to issue Power Purchase Agreements (PPAs).

Influencing provincial and district governments to act to develop and implement energy policy is an enormous challenge for the central government (even more so when issues cannot be effectively addressed within a single region and require negotiation and cooperation between regions).[76] The best means of doing so, predictably, would be through the central government's purse strings, either by providing a special incentive for moving in the right direction or by withholding all central government investment until agreement is reached. Money, however, in and of itself may not be enough. For example, it has also been suggested that "provincial and district governments will need guidance and resources from the central government to develop the capacity to tender and monitor the exploration and exploitation of geothermal working areas" (Girianna 2009: 4). Proposals for financial facilities and technical assistance to support provincial and district governments with tendering, exploitation and monitoring, however, are still on the drawing board. The same applies to regional incentive mechanisms (Ministry of Finance 2009: 12).[77]

A third obstacle lies in the capital cost of developing geothermal technology. Many reports indicate that the cost of geothermal energy is still prohibitive, which is estimated to go up to US $800 million for a 333 MW power plant (or around US $2.4 million/MW[78]). Fully financing such projects would be completely beyond the capacity of the Indonesian government. The government, in early 2010, launched the Indonesia Green Investment Fund (IGIF),[79] putting US $100 million into the fund. It aims to raise a further US $900 million from foreign governments, including Norway and Australia, as well as through institutional investors. Even with such reserves, the fund may only be able to invest in a single geothermal plant.

Therefore, in spite of the development of geothermal energy requiring substantial private investment, associated risks are so disproportionately concentrated in the early resource development stage that it is deterring private investors from entering the market despite the increasingly attractive business propositions being brought forth by changes in pricing by the government. Drilling is expensive, with no guarantees that anything will be found. Risk is further inflated due to the fact that resource information used as a basis for auctioning and price negotiation (under the previous scheme) is less than reliable, as it is primarily based on surface estimations (GeothermEx 2010: 52).

These risks may still be mitigated, for example, by setting up a revolving trust fund that local governments may tap into for the purposes of financing early

---

[76] For example, some regions are landlocked and have no access to ports. In such cases, where adjacent regions control the nearest port, a toll fee that may be so large as to make any commercial initiative unviable is usually demanded.

[77] See the various options floated in the Green Paper (Ministry of Finance 2009).

[78] Shibaki and Beck estimated the geothermal power direct capital costs (installed capacity) to be US $1.9 million/MW (Shibaki/Beck 2003: 10).

[79] Reuters, "Indonesia Readies $1b 'Green' Fund to Spur Clean Investment, Reduce Emission", in: *The Jakarta Globe* (27 January 2010), at: http://www.thejakartaglobe.com/home/indonesia-plans-1-billion-green-investment-fund/354921 (9 January 2013).

resource development activities such as enhancing resource data, with repayments to the fund coming from a share of the income that provincial and district governments would receive from the exploitation of geothermal concessions. Among the initiatives proposed and appear to be in advanced stages of implementation is the Indonesian Geothermal Fund,[80] a rolling fund that allows district governments to borrow funds at a concessional rate for the purposes of enhancing geothermal data of a working area through exhaustive survey prior to auctioning. In order to support the fund, in 2011, the Indonesian government allocated Rp. 1,236.5 billion (or equivalent to US $145 million at the time) (Wahjosoedibjo/Hasan 2012: 1). However, implementation has been stalled due to unresolved issues related to risk sharing and other legal matters pertaining to the management of state funds.

Public private partnership (PPP) arrangements involving international aid agencies are playing their role in the provision of infrastructure. Recognising the high risks of investment in geothermal energy, the World Bank announced a US $400 million commitment from their Clean Technology Fund (CTF) in early 2010.[81] The project is aimed at doubling Indonesia's geothermal energy capacity.[82] In addition, the Lahendong Geothermal Plant is being financed by the World Bank, Asian Development Bank (ADB)[83] and Japan International Cooperation Agency (JICA). [84, 85]

---

[80]   Putri, Adhityani, "Seeking an Affordable Geothermal Energy Price", in: *The Jakarta Post* (27 July 2012), at: http://www.thejakartapost.com/news/2012/07/27/seeking-affordable-geothermal-energy-price.html (9 January 2013).

[81]   "Follow the Money: $400 Million Indonesian Commitment has Players Scurrying", in: *Geothermal Digest* (29 March 2010), at: http://geothermaldigest.net/blog/2010/03/29/follow-the-money-400-million-indonesia-commitment-has-players-scurrying (16 March 2011).

[82]   Padden, Brian, "World Bank Invests $400 Million in Indonesian Geothermal Energy", in: *Voice of America* (23 March 2010), at: http://www.voanews.com/english/news/World-Bank-Invests-400-Million-in-Indonesian-Geothermal-Energy–88906002.html (16 March 2011).

[83]   The Asian Development Bank (ADB) financed the plant as one of 12 subprojects under its Renewable Energy Development Sector (REDS) project, aiming to increase Lahendong's electricity output to 158 gigawatts hour annually into PLN's Minhasa system of North Sulawesi. See World Bank, "ID-PCF-Indonesia Lahendong Geothermal Project" (25 August 2009), at: http://web.worldbank.org/external/projects/main?pagePK=64283627&piPK=64290415&theSite PK=40941&menuPK=228424&Projectid=P096677 (16 March 2011).

[84]   Japan International Cooperation Agency's (JICA) contribution of ¥5,866 million (approximately US $70 million) beginning in March 2004 involved building a new plant with a 20 MW capacity that was due for completion in 2012. See JICA, "Major Projects: Lahendong Geothermal Power Plant Project" (2004), at: http://www.jica.go.jp/indonesia/english/activities/activity13.html (16 March 2011).

[85]   If we use the aforementioned figure of US $800 million for a 333 MW power plant as a guide to the cost of developing geothermal energy in Indonesia, it is reasonably clear that international aid agencies alone will not be able to do much in helping Indonesia shift its energy mix to geothermal. Also, keeping in mind that the global financial crisis is becoming a lingering phenomenon that is continuing to dog most Organisation for Economic Co-operation and Development (OECD) countries, and probably Indonesia as well, it becomes apparent that Indonesia cannot expect much more assistance with its geothermal agenda from international aid sources.

Notwithstanding the government's various initiatives, risk remains exacerbated by an uncertain legal environment. As Chevron President of Asia-Pacific Exploration and Production Jim Blackwell pointed out, Indonesia is unlikely to become a world leader in geothermal until there is "a stable legal and regulatory regime, which allows for long-term development rights, open markets created by long-term contracts and long-term prices with certainty of payment".[86] This quote, however, appears to be inherently contradictory. The quote seems to suggest that private investors would want long-term contracts in the field of geothermal energy at a time when it is likely that long-term gas contracts indexed to the price of oil will be replaced by more flexible markets that reflect the change in gas supply and export competition. One has to further ask whether it would be beneficial for Indonesia to lock itself into long-term geothermal contracts. Another crucial factor that needs to be taken into account in such decision-making is the fact that geothermal energy is locked in particular geographical settings.

Finally, the commercialisation of geothermal energy, in some circumstances, involves a direct conflict with the long-established, well-organised and powerful forestry industry, as an estimated 60 percent of geothermal energy sources are located in forestry areas and also subject to recently enacted laws on pristine forests (including stricter conditions under which licences will be issued) (Girianna 2009: 2). The problem is compounded by the fact that about half of these resources belong to local governments while the rest lie with state-owned companies. In 2009, the then Director General for Renewable Energy and Energy Conservation Luluk Sumiarso publicly guaranteed that the development of geothermal resources would not devastate protected forests since geothermal steam production depended on forest conservation.[87] Reflecting this dichotomy, the possible conflict over forest and land use was given priority by the President in the draft of forest clearance moratorium subsequent to the bilateral agreement with Norway.[88]

The release of the Memorandum of Understanding (MoU) between the MEMR and the Ministry of Forestry No. 7662 of 2011[89] (regarding the acceleration of

---

[86] "Indonesia: $ 5 Billion in Geothermal Deals", in: *United Press International* (28 April 2010), at: http://www.upi.com/Science_News/Resource-Wars/2010/04/28/Indonesia-5-billion-in-geothermal-deals/UPI-75251272464182/ (16 March 2011).

[87] "Geothermal Power Would Not Devastate Forest: Official", in: *The Jakarta Post* (31 March 2011), at: http://www.thejakartapost.com/news/2011/03/31/geothermal-power-would-not-devastate-forests-official.html (16 March 2011).

[88] In the LoI between the Government of Indonesia and the Government of Norway, the two governments agree that taking appropriate measures to address land conflicts is an important part of reducing emissions. See CIFOR, "Letter of Intent between the Government of the Kingdom of Norway and the Government of the Republic of Indonesia on 'Cooperation on Reducing Greenhouse Gas Emissions from Deforestation and Forest Degradation'" (May 2010), at: http://www.forestsclimatechange.org/fileadmin/photos/Norway-Indonesia-LoI.pdf (9 January 2013).

[89] ESDM, "Kementerian Energi Dan Sumber Daya Mineral Republik Indonesia Siaran Pers Nomor: 79/Humas KESDM/2011" (in Bahasa Indonesia), at: http://www.esdm.go.id/siaran-pers/55-siaran-pers/5282-penandatanganan-mou-kementerian-esdm-dengan-kementerian-kehutanan.html (15 July 2013).

permit issuance for geothermal energy development in forest production and forest protection, and preparation for geothermal utilisation in forest conservation areas) has been seen as a breakthrough. This means that the government wishes to fast forward geothermal energy development, even when it overlaps with forest areas. This MoU, however, may be in conflict with higher regulations. Law No. 27 on Geothermal of 2003, for instance, categorises geothermal activities as mining activities. A consequence of this classification is that geothermal energy development cannot take place in forest protection and conservation areas, and is illegal as per the stipulations of Law No. 41 of 1999 on Forestry.[90] It is therefore important to have a process put in place to synchronise such conflicting regulations as well as to institute a set of sustainability benchmarks—as requirement stipulated by the law—that will ensure the mitigation of impacts and risks of geothermal energy development on forests. In order for geothermal energy development to progress more quickly, the government should perhaps consider a 'crash' programme exclusively dedicated to it.

4.4 Further Steps and Conclusion

As discussed in the previous sections, there are a number of obstacles to the successful implementation of Indonesia's climate and energy commitments. Some of these such as the difficulty in removing existing perverse economic incentives and the division of power between the central and provincial governments have been identified. There are also initiatives that could serve as a breakthrough for the issues faced by Indonesian companies involved in developing its renewable energy.

Creating the correct tariff structures for renewable energy and finding ways to finance the development of renewable energy technologies are key priorities that would ensure the continuation and further growth of renewable energy development in the country. Resolving the long battle over fossil fuel subsidies is the government's most important political priority in the energy sector for 2013. Regardless of the amount of funds provided for renewables, the existing policy that keeps in place heavy fuel and electricity subsidies would prove to be counterproductive for both ensuring energy security and climate change mitigation.

This apart, perhaps the greatest difficulties that Indonesia faces are the structural changes taking place in global fossil fuel markets as a result of the revolution underway in extraction technologies. North America is set to become a major exporter of cheap gas; China may also be able to develop its shale gas reserves. In other words, the world has a clear path to energy security based on fossil fuels albeit at the cost of high climate insecurity.

---

[90] See "Law of the Republic of Indonesia Number 41 of 1999 Regarding Forestry", at: http://www.theredddesk.org/sites/default/files/uu41_99_en.pdf (15 July 2013).

Indonesia cannot develop its geothermal resources without large-scale foreign investment. As we have seen, foreign investment to date has been modest and much of it has come from international aid agencies. Indonesia could offer foreign investors the security of premium long-term contracts to develop its geothermal reserves, but one has to ask whether this would be a rational strategy in a world of changing gas, oil and coal markets and cheaper prices. Should Indonesia proceed to pursue such options, other countries would benefit from competitive energy prices while Indonesia would be locked into its expensive geothermal sources. What is more, such unilateralism on geothermal energy on the part of Indonesia would not mitigate climate insecurity.

Indonesia's current strategy of muddling through on its geothermal options may be the right one, at least until the consequences of the structural transformations currently taking place in the gas and oil markets become clearer. Arguably, Indonesia has done enough on climate change and energy for a developing country with a large population and many millions of poor people. It cannot afford to lock itself into expensive geothermal options and paying monopoly rents to foreign private investors in geothermal while other countries gain competitive advantages from falling fossil fuel prices. It can, as it is doing presently, develop a regulatory framework for geothermal energy and take advantage of international aid money to develop some geothermal plants. As we have indicated, it will take billions of dollars to fully develop the geothermal resources in Indonesia and all that is on the table at the moment are a few hundred million dollars. What the Indonesian President could do, in the meantime, is point out to Organisation for Economic Co-operation and Development (OECD) countries that if they decide on choosing fossil fuel energy security and live with deep climate insecurity this would be their choice; under the present circumstances, Indonesia can do no more than what it is doing presently.

References

Ardiansyah, Fitrian, 2010: "The Roller-coaster of Indonesia's Leadership in Climate Negotiations", An International Affairs Forum Special Report (Winter 2009/2010): 56–61.

Ardiansyah, Fitrian; Gunningham, Neil; Drahos, Peter, 2012: "An Environmental Perspective on Energy Development in Indonesia", in: Caballero-Anthony, Mely; Chang, Youngho; Putra, Nur Azha (Eds.): *Energy and Non-Traditional Security (NTS) in Asia* (Heidelberg: Springer): 89–117.

Ariati, Ratna, 2009: "Kebutuhan Mitigasi Perubahan Iklim di Sektor Energi dan Pendanaannya (The Needs for Climate Change Mitigation in the Energy Sector and Its Financing Options)", Paper for the NEEDS for Climate Change Launching, Jakarta, Indonesia.

Asafu-Adjaye, John, 2000: "The Relationship between Energy Consumption, Energy Prices and Economic Growth: Time Series Evidence from Asian Developing Countries", in: *Energy Economics*, 22,6: 615–625.

Beaton, Christopher; Lontoh, Lucky, 2010: *Lessons Learned from Indonesia's Attempts to Reform Fossil-Fuel Subsidies* (Manitoba: International Institute for Sustainable Development).

Bruch, Michael; Münch, Volker; Aichinger, Markus; Kuhn, Michael; Weymann, Martin; Schmid, Gerhard, 2011: "Power Blackout Risks: Risk Management Options", Emerging Risk Initiative—Position Paper, November 2011 (Amsterdam: CRO Forum).

Castlerock, 2010: *Phase 1 Report: Review and Analysis of Prevailing Geothermal Policies, Regulations and Costs*, A Report Commissioned by ESDM (MEMR) (Jakarta: ESDM).

Darma, Surya; Harsoprayitno, Sugiharto; Setiawan, Bambang; Hadyanto; Sukhyar, R.; Soedibjo, Anton W.; Ganefianto, Novi; Stimac, Jim, 2010: "Geothermal Energy Update: Geothermal Energy Development and Utilization in Indonesia", Paper for World Geothermal Congress 2010, Bali, Indonesia, 25–29 April.

Desai, Raj M.; Olofsgård, Anders; Yousef, Tarik M., 2009: "The Logic of Authoritarian Bargains", in: *Economics and Politics*, 21,1: 93–125.

DNPI (Dewan Nasional Perubahan Iklim [National Council on Climate Change]), 2010: *Indonesia's Greenhouse Gas Abatement Curve* (Jakarta: DNPI).

GeothermEx, 2010: *An Assessment of Geothermal Resource Risks in Indonesia* (Washington, D.C.: PPIAF-World Bank).

Girianna, Montty, 2009: "Renewable Energy and Energy Efficiency in Indonesia", Paper for the ADB Workshop on Climate Change and Energy, Bangkok, Thailand, 26–27 March.

Hilmawan, Edi; Said, Mustafa, 2009: "Energy Efficiency Standard and Labelling Policy in Indonesia", Paper for the International Cooperation for Energy Efficiency Standard and Labeling Policy Conference, Tokyo, Japan.

Howarth, Robert W; Santoro, Renee; Ingraffea, Anthony, 2011: "Methane and the Greenhouse-gas Footprint of Natural Gas from Shale Formations: A Letter", in: *Climatic Change*, 106,4: 679–690.

IEA (International Energy Agency), 2012: *World Energy Outlook 2012* (Paris: IEA).

Jafari, Yaghoob; Othman, Jamal; Nor, Abu Hassan Shaari Mohd, 2012: "Energy Consumption, Economic Growth and Environmental Pollutants in Indonesia", in: *Journal of Policy Modeling*, 34,6: 879–889.

Jotzo, Frank, 2012; "Can Indonesia Lead on Climate Change?", in: Reid, Anthony J.S. (Ed.): *Indonesia Rising: The Repositioning of Asia's Third Giant* (Singapore: Institute of Southeast Asian Studies): 92–115.

Leitmann, Josef, et al., 2009: *Investing in a More Sustainable Indonesia: Country Environmental Analysis*, Country Environmental Analysis 2009, Report No. 50762 (Jakarta: World Bank).

Mackay, David, 2008: *Sustainable Energy: Without the Hot Air* (Cambridge: UIT).

Marpaung, C.O.P.; Widodo, Bambang; Soebagio, Atmonobudi; Purba, R.; Ambarita, E., 2012: "Energy Security Implications of Introducing Renewable Portfolio Standard in Indonesia", Paper for the 2nd Congress of the East Asian Association of Environmental and Resource Economics, Bandung, Indonesia, 2–4 February.

Melisa, Eka, 2010: "National Action Plan and International Partnership: Indonesia's Response to Global Efforts to Improve the International Regime on Climate Change", Paper for Asia Climate Change Policy Forum 2010, Canberra, Australia, 27 October.

Ministry of Environment, 2010: *Indonesia Second National Communication Under the United Nations Framework Convention on Climate Change* (Jakarta: Ministry of Environment).

Ministry of Finance, 2009: *Ministry of Finance Green Paper: Economic and Fiscal Policy Strategies for Climate Change Mitigation in Indonesia* (Jakarta: Ministry of Finance and Australia Indonesia Partnership).

Nurdianto, Ditya A.; Resosudarmo, Budy P., 2011: "Prospects and Challenges for an ASEAN Energy Integration Policy", in: *Environmental Economics and Policy Studies*, 13,2: 103–127.

OECD (Organisation for Economic Co-operation and Development); IEA (International Energy Agency), 2012: *Golden Rule for a Golden Age of Gas: World Energy Outlook Special Report on Unconventional Gas* (Paris: OECD/IEA).

Pertamina; WLPGA (World Liquid Petroleum Gas Association), 2012: *Kerosene to LP Gas Conversion Programme: A Case Study of Domestic Energy* (Jakarta: Pertamina and WLPGA).

Pozon, Ina; Ashat, Ali; Ardiansyah, Fitrian, 2012: *Igniting the Ring of Fire: A Vision for Developing Indonesia's Geothermal Power* (Jakarta: WWF-Indonesia).

Purnomo, Agus; Katili-Niode, Amanda; Melisa, Eka; Helmy, Farhan; Sukadri, Doddy; Sitorus, Suzanty, 2013: *Evolution of Indonesia's Climate Change Policy: From Bali to Durban* (Jakarta: Dewan Nasional Perubahan Iklim).

PwC, 2011: *Electricity in Indonesia: Investment and Taxation Guide* (Jakarta: PwC Indonesia).

PwC, 2012: *Oil and Gas in Indonesia: Investment and Taxation Guide* (Jakarta: PwC Indonesia).

Resosudarmo, Budy P.; Alisjahbana, Armida; Nurdianto, Ditya A., 2010: "Energy Security in Indonesia", ANU Working Paper in Trade and Development (Canberra: Crawford School, ANU).

Sa'ad, Suleiman, 2009: "An Empirical Analysis of Petroleum Demand for Indonesia: An Application of the Cointegration Approach", in: *Energy Policy*, 37,11: 4391–4396.

Shibaki, Mashashi; Beck, Fredric, 2003: "Geothermal Energy for Electric Power", A REPP Issue Brief, December 2003 (Washington, D.C.: Renewable Energy Policy Project).

Sukhyar, R., 2011: "Pengembangan Panas Bumi di Indonesia: Menanti Pembuktian" (Geothermal Development in Indonesia: Waiting for Realisation), Paper for National Seminar on Geothermal: Our Savior for a Better Tomorrow, Bandung, Indonesia, 12 February.

Wahjosoedibjo, Anton; Hasan, Madjedi, 2012: "Geothermal Fund for the Hastening of the Development of Indonesia Geothermal Resources", Paper for the Thirty-Seventh Workshop on Geothermal Reservoir Engineering, Stanford University, Stanford, California, United States, 30 January–1 February.

Wijaya, Muhammad Ery; Limmeechokchai, Bundit, 2009: "Optimization of Indonesian Geothermal Energy Resources for Future Clean Electricity Supply: A Case of Java-Madura-Bali System", in: *IIRE International Journal of Renewable Energy*, 4,2: 1–12.

World Bank, 2008: *Geothermal Power Generation Development Project* (Washington, D.C.: World Bank).

Abbreviations

ADB	Asian Development Bank
AIM	Asian-Pacific Integrated Model
ANU	Australian National University
APBN-P	Anggaran Pendapatan dan Belanja Negara–Perubahan (The Revised Government Budget and Expenditure)
APEC	Asia–Pacific Economic Cooperation
BAP	Bali Action Plan
BAPPENAS	Badan Perencanaan Pembangunan Nasional (National Development Planning Agency)
BAU	'Business as usual'
BIICL	British Institute of International and Comparative Law
CARR	Centre for the Analysis of Risk and Regulation
CIFOR	Center for International Forestry Research
CO_2	Carbon dioxide
CO_2e	CO_2 equivalent
COP	Conference of the Parties
CTF	Clean Technology Fund
DEN	Dewan Energy Nasional (National Energy Council)
DG-EBTKE	Directorate General of New, Renewable Energy, and Energy Conservation
DNPI	Dewan Nasional Perubahan Iklim Indonesia (National Council on Climate Change)

EAAERE	East Asian Association of Environmental and Resource Economics
EIA	U.S. Energy Information Administration
ESCO	Energy Services Company
ESDM	Kementerian Energi dan Sumber daya Mineral (Ministry of Energy and Mineral Resources, Republic of Indonesia)
FAO	Food and Agriculture Organization of the United Nations
FAOLEX	Legislative database of FAO Legal Office
FIT	Feed-in-tariff
GHG	Greenhouse gas
GW	Gigawatt
ICCTF	Indonesia Climate Change Trust Fund
IEA	International Energy Agency
IFCA	Indonesia Forest Climate Alliance
IGIF	Indonesia Green Investment Fund
JICA	Japan International Cooperation Agency
kW	Kilowatt
kWh	Kilowatt hour
LoI	Letter of intent
MEMR	Ministry of Energy and Mineral Resources
MoU	Memorandum of Understanding
Mt	Million tonnes
MW	Megawatt
MWe	Megawatts-electrical
NTS	Non-traditional security
OECD	Organisation for Economic Co-operation and Development
OHS	Occupational health and safety
PERTAMINA	Perusahaan Tambang dan Minyak Negara (state-owned oil and gas company)
PLN	Perusahaan Listrik Negara
PPN	Perencanaan Pembangunan Nasional (National Development Planning)
PPP	Public private partnership
PPA	Power purchase agreements
PPIAF	Public-Private Infrastructure Advisory Facility
PwC	PricewaterhouseCoopers
RAN-GRK	Rencana Aksi Nasional Penurunan Emisi Gas Rumah Kaca: (National Action Plan on GHG Emissions Reduction)
REDD+	Reducing Emissions from Deforestation and Forest Degradation-Plus
REDS	Renewable Energy Development Sector
RegNet	Regulatory institutions network
REPP	Renewable Energy Policy Project
RPS	Renewable portfolio standard

RSPO	Roundtable on Sustainable Palm Oil
SME	Small and medium enterprise
Tbd	Thousand barrels per day
UNFCCC	United Nations Framework Convention on Climate Change
WCA	World Coal Association
WLPGA	World Liquid Petroleum Gas Association
WWF	World Wide Fund for Nature

Chapter 5
Singapore's Policy Response to Climate Change: Towards a Sustainable Future

Nur Azha Putra and Nicholas Koh

Abstract This chapter examines the sustainability of Singapore's climate change policies. It argues that while the state has demonstrated its political will to reduce its carbon footprint, adapt to the effects of climate change and mitigate the country's long-term vulnerabilities, the nation as a whole needs to build upon this work towards a more profound solution by nurturing a sustainable Singaporean society.

5.1 Introduction

Singapore has a history of climate change awareness. Singapore ratified the United Nations Framework Convention on Climate Change (UNFCCC) in 1997[1] and acceded to the Kyoto Protocol of the UNFCCC on 12 April 2006.[2] In the same month, the Climate Change Awareness Programme (CCAP)[3] was launched by the Singapore Environment Council (SEC),[4] which was supported by the National

---

[1] NCCS, "Singapore's International Actions", April 2012, at: http://app.nccs.gov.sg/page.aspx?pageid=55&AspxAutoDetectCookieSupport=1 (12 June 2013).
[2] Ibid.
[3] NEA, "Everyday Superhero Saves the Planet", 2005–2006, at: http://www.nea.gov.sg/ar06/02SideSuperHero.html (12 June 2013).
[4] Ibid.

N. Azha Putra (✉)
Energy Studies Institute (ESI), National University of Singapore (NUS), 29 Heng Mui Keng Terrace, Block A, #10-01, Singapore 119620, Singapore
e-mail: azha@nus.edu.sg
URL: http://www.esi.nus.edu.sg/about-us/our-researchers/nur-azha-putra

N. Koh
National Research Foundation (NRF), 1 Create Way, #12-02, Singapore 138602, Singapore
e-mail: nickkoh@gmail.com

N. A. Putra and E. Han (eds.), *Governments' Responses to Climate Change: Selected Examples From Asia Pacific* 10, SpringerBriefs in Environment, Security, Development and Peace, DOI: 10.1007/978-981-4451-12-3_5, © The Author(s) 2014

Environment Agency (NEA). The aim of the year-long programme was to raise awareness among Singaporeans about climate change. Other activities included the release of the National Climate Change Strategy in 2008[5] and 2012,[6] which were a set of speeches and statements published by the Ministry of the Environment and Water Resources (MEWR) and the NEA. On 12 March 2010, Singapore's National Climate Change Secretariat (NCCS) was placed under the Prime Minister's Office, headed by a Permanent Secretary.[7] Earlier, an Inter-Ministerial Committee on Climate Change (IMCCC)[8] was set up in 2007 to oversee inter-agency coordination while implementing government directives for addressing climate change.

A host of national policy reports were also released to educate the general public, acknowledge the importance of climate change and outline national strategies and policy recommendations.[9] This included the Singapore Green Plan 2012 published in February 2006,[10] the National Energy Policy Report published in November 2007,[11] the National Climate Change Strategy published in March 2008[12] and its updated version in 2012,[13] the Sustainable Development Blueprint published in April 2009,[14] and Singapore's Second National Communication on Climate Change published in November 2010.[15]

---

[5] MEWR, "Singapore's National Climate Change Strategy", March 2008, at: http://www.elaw.org/system/files/Singapore_Full_Version.pdf (1 July 2013).

[6] NCCS, "National Climate Change Strategy 2012", June 2012, at: http://app.nccs.gov.sg/data/resources/docs/Documents/NCCS-2012.pdf (12 June 2013).

[7] "National Climate Change Secretariat to Come under PMO from July 1", in: *Xin MSN News* (25 June 2010), at: http://news.xin.msn.com/en/singapore/article.aspx?cp-documentid=4175806 (12 July 2013).

[8] NCCS, "Inter-Ministerial Committee on Climate Change", April 2013, at: http://app.nccs.gov.sg/page.aspx?pageid=47 (12 June 2013).

[9] MEWR, "Grab Our Research", June 2013, at: http://app.mewr.gov.sg/web/Contents/Contents.aspx?Id=195 (12 June 2013).

[10] MEWR, "The Singapore Green Plan 2012", February 2006, at: http://app.mewr.gov.sg/data/ImgCont/1342/sgp2012_2006edition.pdf (1 July 2013).

[11] MTI, "Energy for Growth: National Energy Policy Report", November 2007, at: http://www.mti.gov.sg/ResearchRoom/Documents/app.mti.gov.sg/data/pages/885/doc/NEPR%202007.pdf (1 July 2013).

[12] MEWR, "Singapore's National Climate Change Strategy", March 2008, at: http://www.elaw.org/system/files/Singapore_Full_Version.pdf (1 July 2013).

[13] NCCS, "Climate Change & Singapore: Challenges, Opportunities, Partnerships", National Climate Change Strategy 2012, at: http://app.nccs.gov.sg/data/resources/docs/Documents/NCCS-2012.pdf (20 January 2013).

[14] MEWR; MND, "A Lively and Liveable Singapore: Strategies for Sustainable Growth", at: http://app.mewr.gov.sg/data/ImgCont/1292/sustainbleblueprint_forweb.pdf (20 January 2013).

[15] NEA, "Singapore's Second National Communication, Under the United Nations Framework Convention on Climate Change", November 2010, at: http://app.nccs.gov.sg/data/resources/docs/SINGAPORE%27S%20SECOND%20NATIONAL%20COMMUNICATIONS%20NOV%202010.pdf (1 July 2013).

5.2 Singapore's Strategies to Address Climate Change

Singapore's efforts to mitigate and adapt to climate change can broadly be cate-gorised as: (a) increasing the country's energy efficiency; (b) utilising fuels that emit less carbon dioxide (CO_2); (c) enhancing social resilience; (d) promoting green technology; and, (e) participating in international regimes on climate change.[16] These are in addition to early measures such as reducing the growth of its vehicle population and promoting the optimal use of scarce land through integrated urban planning.[17]

5.2.1 Mitigation

5.2.1.1 Reducing Emissions

Singapore is a small city-state with a high population density. She is a major refining and petrochemicals hub, with the industry situated on Jurong Island, which is an island south-west off the main island of Singapore. Along with electricity generation, transportation, buildings and households, the industrial sector is a major contributor to Singapore's total CO_2 emissions although the products it manufactures are mainly for export purposes.

The overall figures for primary and secondary consumption[18] for Singapore reveal that, in 2005, the industries, which emitted a total of 21,793 kilo tonnes of CO_2, accounted for 54 percent of the total CO_2 emitted in the country (Table 5.1).[19]

According to the International Energy Agency (IEA) though, Singapore's CO_2 emissions, which were at 62.9 million tonnes in 2010, are well behind other devel-oped economies.[20] The countries that were the top ten CO_2 emitters in 2010 were China, the United States, India, the Russian Federation, Japan, Germany, Korea, Canada, Islamic Republic of Iran and the United Kingdom.[21] These countries emitted a total 19.8 Gt of CO_2 in 2010 while the global CO_2 emissions were 30.3 Gt.[22]

---

[16] NCCS, "Climate Change & Singapore: Challenges, Opportunities, Partnerships", National Climate Change Strategy 2012, at: http://app.nccs.gov.sg/data/resources/docs/Documents/NCCS-2012.pdf (20 January 2013): 35.

[17] Ibid.

[18] Primary consumption figures indicate consuming sectors that generate energy directly from fuel combustion while secondary consumption figures are indicative of sectors that utilise electricity generated from fuel combustion.

[19] MEWR, "Singapore's National Climate Change Strategy", March 2008, at: http://www.elaw.org/system/files/Singapore_Full_Version.pdf (1 July 2013).

[20] See IEA, "CO_2 Emissions From Fuel Combustion: Highlights", at: http://www.iea.org/co2highlights/co2highlights.pdf (24 January 2012): 50.

[21] Ibid: 9.

[22] Ibid.

Table 5.1 Sectors contributing to Singapore's CO_2 emissions in 2005

	Contributing sectors (kilo tonnes of CO_2 (%))					
	Electricity generation	Industry	Transport	Buildings	Consumers/ households	Other
Primary consumption (combust fuel)	19,315 (48)	13,465 (33)	7,056 (17)	325 (1)	216 (1)	–
Secondary consumption (use electricity)		8,328 (21)	930 (2)	5,910 (15)	3,415 (8)	732 (2)
Overall		21,793 (54)	7,986 (19)	6,235 (16)	3,631 (9)	732 (2)

Source Adapted from MEWR 2008: MEWR, "Singapore's National Climate Change Strategy", March 2008, at: http://www.elaw.org/system/files/Singapore_Full_Version.pdf (1 July 2013) 15, at: http://www.elaw.org/system/files/Singapore_Full_Version.pdf(1 July 2013)

5.2.1.2 Increasing Energy Efficiency

One of the key thrusts of Singapore's climate change mitigation strategy has been to improve energy efficiency. Within Singapore, efforts towards energy efficiency are consolidated by way of a partnership between various government statutory boards, with the NEA and the Energy Market Authority (EMA) chairing the Energy Efficiency Singapore Programme Office (E^2PO). E^2PO members include the Economic Development Board (EDB), Land Transport Authority (LTA), Building and Construction Authority (BCA), Housing and Development Board (HDB), Infocomm Authority of Singapore (IDA), Agency for Science, Technology and Research (A*STAR), Urban Redevelopment Authority (URA), Jurong Town Corporation (JTC) and National Research Foundation (NRF).[23] In October 2009, a new voluntary partnership programme—Energy Efficiency National Partnership (EENP)—was launched by the NEA to target the improvement of energy efficiency within the industrial sector.[24]

According to the EMA, households accounted for 15.7 percent of the total national electricity consumption in Singapore in 2011.[25] In the same year, the most energy-intensive sectors were the industries (40.2 percent) and the commerce- and

---

[23] E^2PO, "About E^2PO: Objective and Members", at: http://app.e2singapore.gov.sg/About_Esup2/supPO/Objective_and_Members.aspx (1 July 2013).

[24] NEA, "National Environment Agency Launches A New Voluntary Partnership Programme To Promote Energy Efficiency In Industry" (30 October 2009), at: http://app2.nea.gov.sg/corporate-functions/newsroom/news-releases/year/2009/month/10/category/nea-news-release/national-environment-agency-launches-a-new-voluntary-partnership-programme-to-promote-energy-efficiency-in-industry (1 July 2013).

[25] EMA, "Energising Our Nation: Singapore Energy Statistics 2012", at: http://www.ema.gov.sg/media/files/publications/EMA_SES_2012_Final.pdf (20 January 2013): 18.

services-related sectors (37.5 percent).[26] The transport sector meanwhile only accounted for 5.5 percent of electricity sales.[27]

Singapore's energy intensity has been steadily declining from its peak in 1991.[28] According to the NEA, its energy intensity improved by 15 percent between 1990 and 2005 due to the adoption of better technology in power generation and the more productive use of energy in other sectors.[29] The NEA reported also that Singapore's energy intensity improved by 16 percent between 2005 and 2010.[30]

5.2.1.3 Altering the Fuel Mix: From Oil to Gas

To reduce the usage of petrol for electricity generation, Singapore has since the early 2000s shifted towards a predominantly natural gas energy mix.[31] As of 2011, piped natural gas accounted for 78 percent of Singapore's fuel mix for electricity generation, with petroleum products accounting for 18.4 percent of the fuel mix and the remaining 3.6 percent comprised of fuel types such as synthetic gas, diesel and municipal wastes.[32]

Singapore started using natural gas in the early 1990s, with consumption increasing by nearly eight times to 8.1 billion cubic metres in 2009.[33] Natural gas has not only overtaken oil as the dominant fuel for electricity generation in the country but is also widely used in the petrochemical sector. Currently, Singapore's gas is entirely supplied via pipelines from Indonesia and Malaysia. This scenario is set to change with the completion of a new US $1.7 billion liquefied natural gas (LNG) terminal in 2013,[34] which will allow Singapore to import LNG via ships. This development will help Singapore enhance its energy security by diversifying its gas

---

[26] Ibid.

[27] Ibid.

[28] NEA, "E² Singapore", at: http://app.e2singapore.gov.sg/DATA/0/docs/Booklet/E2S%20Publication.pdf (18 June 2011): 5.

[29] Ibid.

[30] "About E²PO: Energy Efficiency in Singapore", at: http://app.e2singapore.gov.sg/About_Esup2/supPO/Why_Energy_Efficiency/Climate_Change_and_Energy_Efficiency.aspx (29 Feb 2013).

[31] EMA, "Fuel Mix for Electricity Generation (2006–2011)", at: http://www.ema.gov.sg/media/files/publications/EMA_SES_2012_Final.pdf (14 January 2012).

[32] Ibid.

[33] BP, "BP Statistical Review of World Energy June 2011", at: http://www.bp.com/assets/bp_internet/globalbp/globalbp_uk_english/reports_and_publications/statistical_energy_review_2011/STAGING/local_assets/pdf/statistical_review_of_world_energy_full_report_2011.pdf (1 July 2013): 23.

[34] SLNG, "About the Singapore LNG Terminal", at: http://www.slngcorp.com/about-us-lng-terminal.html (14 December 2011).

sources,[35] especially as a contingency plan should Indonesia and Malaysia divert gas meant for exports for domestic usage. Reports that Indonesia has plans to reduce natural gas exports to Singapore by 2020[36] only serve to underscore the importance of Singapore having alternative sources of natural gas via its LNG terminal.

The LNG terminal will have two storage tanks, with a capacity of 180,000 cubic metres each.[37]. The terminal and the two tanks are to be completed by 2013, following which a third tank with additional storage capacity of 180,000 cubic metres will be added by 2014.[38] Singapore's first shipment of LNG was supplied by QatarGas Operating Company Limited (QatarGas), which is the largest LNG producer in the world.[39] While the bulk of LNG would be used to generate power in Singapore, other possible uses for LNG ranging from cryogenic power generation from cold energy to manufacturing industrial gases and liquefied petroleum gas trading are being explored.[40]

Other Southeast Asian countries, notably Vietnam and Thailand, are also in the midst of building gasification terminals, to be ready over the next few years, as they diversify and move away from more polluting fossil fuels.[41] This will add to the already increasing demand for LNG from existing heavy LNG users such as Japan, Korea and Taiwan as well as future major consumers such as India and China. However, it remains to be seen if the increase in regional demand will affect Singapore significantly, either in terms of pricing or the availability of LNG. With Australia expected to become a major LNG player in the future,[42] Iran and Qatar in the Middle East ramping up production, and the possibility of obtaining

---

[35] MTI; EMA; SLNG, "LNG Terminal will Diversify Energy Sources and Enhance Singapore's Energy Security", at: http://www.slngcorp.com/UserFiles/Press/FINAL_Media_Release_on_LNG_visit_13_Feb_2012_website.pdf (14 December 2012).

[36] "Indonesia May Seek to Reduce Natural Gas Exports to Singapore—Minister", in: *Dow Jones Newswire* (17 June 2010), at: http://www.nasdaq.com/aspx/stock-market-newsstory.aspx?storyid=201006162112dowjonesdjonline000767&title=indonesia-may-seek-to-reduce-natural-gas-exports-to-singapore-minister#ixzz0r4jZpLis (14 July 2011).

[37] See footnote 34.

[38] EMA, "Third Tank for Singapore's LNG Terminal on the Back of Strong LNG Uptake" (2 November 2010), at: http://www.ema.gov.sg/news/view/227 (1 July 2013).

[39] SLNG, "Singapore's First LNG Shipment to Come from Qatar" (29 November 2012), at: http://www.slng.com.sg/UserFiles/Press/20121129-commissioningcargo.pdf (14 December 2012).

[40] "SLNG Eyes Business Spinoffs from $1.5b Gas Terminal", in: *The Business Times* (17 June 2010), at: http://www.businesstimes.com.sg/sub/news/story/0,4574,390808,00.html (14 July 2011).

[41] "SE Asia to Start LNG Imports in Next 3 Yrs", in: *Reuters News* (2 June 2010), at: http://www.reuters.com/article/idUSSGE6510AV20100602 (14 July 2011).

[42] Australia has three major natural gas projects under construction—Gorgon, Browse and Bonaparte—with access to total estimated reserves of 2.07 trillion cubic metres (tcm). The Gorgon gas field by itself has 1.2 tcm of natural gas reserves (original figure was 68.9 tcf). See EIA, "Country Analysis Briefs: Australia Energy Data, Statistics and Analysis—Oil, Gas, Electricity, Coal", January 2007, at: http://www.eia.doe.gov/emeu/cabs/Australia/Full.html (14 July 2011).

LNG shipments from distant countries such as Egypt and Trinidad & Tobago on the horizon, there appear to exist ample supplies that could satisfy Singapore's demand for gas over the next 20 years.

5.2.1.4 Competency Building

In 2013, Singapore introduced an Energy Conservation Act that would enforce minimum energy requirements for industry players.[43] The Act sets out to achieve a 35 percent improvement in Singapore's energy intensity by 2030 from 2005 levels, improve the energy performance of companies, encourage the industrial sector to further its capacity building and invest in energy efficiency technologies, and standardise energy efficiency standards across all sectors.[44]

Several national programmes aimed at the industries are in place to assist the sector achieve the objectives of the Act. One such programme is the Singapore Certified Energy Managers (SCEM) Training Grant,[45] which subsidises companies that are training their employees to be energy managers. Trained energy managers would have the necessary skills to look into their companies' energy requirements, report energy usage and propose energy efficiency improvement plans. With time, the industries would be a self-regulating entity capable of managing its energy needs efficiently.

Additionally, to advance the country's research and development (R&D) capabilities in clean and efficient energy, Singapore has established several technology research institutes such as the Solar Energy Research Institute of Singapore (SERIS), which was officially launched on 1 April 2008.[46] SERIS, an autonomous organisation under the National University of Singapore (NUS), is focused on solar energy research. Similarly, the Energy Research Institute at Nanyang Technological University (NTU) [ERI@N], launched on 15 June 2010,[47] focuses on "areas of sustainable energy, energy efficiency/infrastructure, and socio-economic aspects of energy research".[48] Meanwhile, the Energy Studies Institute (ESI) at NUS was launched in 2007.[49] Its areas of research focus on energy

---

[43] EMA, "A New Act Comes into Play", in: *On* (July 2012), at: http://www.ema.gov.sg/news/newsletters/2012/07/insighton-newact.html (24 January 2013).

[44] EMA, "Factsheet: Energy Conservation Act", at: http://app.mewr.gov.sg/data/ImgCont/1386/2.%20Factsheet_Energy%20Conservation%20Act%20%5Bweb%5D.pdf (24 January 2013).

[45] SEI, "Singapore Certified Energy Manager (Professional Level)", at: http://www.nea.gov.sg/cms/sei/SCEM.html (25 April 2013).

[46] SERIS, "Solar Energy Research Institute (SERIS) Annual Report 2010", at: http://www.seris.sg/site/servlet/linkableblob/main/4492/data/pdf_2011_06_28_full_AR2010_single_page-data.pdf (14 July 2011): 7.

[47] "ERI@N Official Launch", at: http://www3.ntu.edu.sg/erian/opening.html (14 July 2011).

[48] "ERI@N Overview", at: http://www3.ntu.edu.sg/erian/about_us.htm (14 July 2011).

[49] ESI, "Launch of New Energy Studies Institute at NUS", at: http://www.esi.nus.edu.sg/eventitem/2011/10/03/official-launch-of-esi (14 January 2013).

security, energy economics, and climate and the environment. The ESI has since gone on to become a globally ranked energy research institute.[50]

5.2.2 Adaptation: Enhancing Resilience to Climate Change Effects

Singapore's annual mean surface temperatures have increased from 26.8 °C in 1948 to 27.6 °C in 2011, in line with the trends seen in the country's urban density.[51] Between 1980 and 2010, the daily rainfall records also show an upward trend, just as the mean sea level has increased by 3 mm per year.[52] Given such circumstances, Singapore has responded to the climate change challenges before it with various adaptation strategies, as the country needs to adapt to its changing climate and environment. For instance, the government will be increasing its efforts to protect the country's coastlines and improve the overall drainage system—the latter as a strategy to increase the country's resilience against flood risks. Similarly, to adapt to rising sea levels, the minimum reclamation levels for newly reclaimed land was increased by 1 m as of 2011, which is in addition to the previous requirement of 1.25 m.[53] To protect its water security, the national water agency, Public Utilities Board (PUB), will strengthen Singapore's water energy supplies by enhancing the robustness of its local water catchments areas, and using imported water, NEWater[54] and desalinated water.[55] Ongoing studies have also been commissioned by the government to further identify specific climate change threats and the country's long-term vulnerabilities.[56]

The myriad of Singapore's adaptation strategies are not implemented in isolation. They are, in fact, guided by a common rationale and unified by the philosophy of 'resilience'. The Singapore government devised a resilience framework

---

[50] University of Pennsylvania, "2012 Global Go To Think Tanks Report and Policy Advice", at: http://www.gotothinktank.com/wp-content/uploads/2013/01/2012-Global-Go-To-Think-Tank-Report.pdf (30 January 2013): 83.

[51] NCCS, "Climate Change & Singapore: Challenges, Opportunities, Partnerships", National Climate Change Strategy 2012, at: http://app.nccs.gov.sg/data/resources/docs/Documents/NCCS-2012.pdf (14 January 2013): 70.

[52] Ibid.

[53] Ibid.

[54] NEWater is high-grade reclaimed water that is produced from treated used water, further purified using advanced membrane technologies and ultraviolet disinfection, making it ultra clean and safe to drink. See PUB, "NEWater", at: http://www.pub.gov.sg/water/newater/Pages/default.aspx (20 January 2013).

[55] PUB, "Four National Taps Provide Water for All", at: http://www.pub.gov.sg/water/Pages/default.aspx (10 June 2013).

[56] NCCS, "Climate Change & Singapore: Challenges, Opportunities, Partnerships", National Climate Change Strategy 2012, at: http://app.nccs.gov.sg/data/resources/docs/Documents/NCCS-2012.pdf (14 January 2013): 73.

(Fig. 5.1) with the intent of providing its adaptation strategies a coherent approach to the long-term effects of climate change. Its adaptation strategies were based on a deep understanding of climate change and its evolving implications, which in turn were driven by research and scientific studies, and a risk management framework, as highlighted below.

5.2.3 Economic Opportunities in Green Growth

In mitigating and adapting to the threats of climate change, the Singapore government has identified several key areas that could provide opportunities for economic growth. One such area is the possibility of making the country a global clean technology (cleantech) hub. Due to climate change, countries would need to deploy clean and sustainable energy solutions in addition to managing depleting fossil fuel resources and rapid urbanisation. The global clean energy market is expected to grow strong in the future and global investments are expected to grow beyond the estimated US $260 billion in 2011.[57] The government therefore sees strong economic incentives for pursuing R&D in clean technology. To capitalise on the global demand for clean energy, the Singapore government intends to invest in three key areas: (a) technological capabilities in R&D and human capital; (b) providing test-bedding facilities for the testing of new technologies and processes; and, (c) providing a conducive business environment that offers strong intellectual property regimes and a highly trained supporting workforce.[58]

5.2.4 Partnerships in Climate Change Regimes[59]

5.2.4.1 Local Partnerships

Within the country, the government engages non-state actors to promote and use green practices. It does so by raising awareness of climate change among the public and encouraging the adoption of green practices by the private sector. Environmental issues are taught in public schools, and students are brought on site

---

[57] Ibid: 90.

[58] Ibid: 92.

[59] NEA, "Singapore's Second National Communication, Under the United Nations Framework Convention on Climate Change", November 2010, at: http://app.nccs.gov.sg/data/resources/docs/ SINGAPORE%27S%20SECOND%20NATIONAL%20COMMUNICATIONS%20NOV%20 2010.pdf (1 July 2013).

Fig. 5.1 Singapore's resilience framework

visits to power stations, incineration plants and meteorological stations.[60] Beyond schools, there are the occasional national campaigns such as the Clean and Green campaigns that encourage Singaporeans to adopt environmentally friendly lifestyles. National energy labelling schemes too help the population make informed purchases of energy and environmentally friendly household appliances.

---

[60] NCCS, "Climate Change & Singapore: Challenges, Opportunities, Partnerships", National Climate Change Strategy 2012, at: http://app.nccs.gov.sg/data/resources/docs/Documents/NCCS-2012.pdf (20 January 2013): 116.

5.2.4.2 International Partnerships

Singapore is among the countries that have ratified the UNFCCC. As a non-Annex I member country, Singapore has pledged to continuously contribute towards multilateral efforts aimed at mitigating climate change through multilateral cooperation via other UN specialised agencies.[61] Additionally, Singapore engages in bilateral agreements with China, specifically for the Sino-Singapore Tianjin Eco-city project, and Indonesia for promoting sustainable land use practices and management.[62]

5.3 Singapore's Climate Change Policies: Will These be Enough?

A key question to ask at this juncture is whether Singapore's climate change policies would be enough to ensure that the country remains safe from environmental and climatic degradation while also remaining economically competitive in the long run? It appears that Singapore has put in place commendable efforts to protect its physical environment and ensure that it maximises the utility of its energy resources in a sustainable manner.

The government intends to reduce energy intensity in households through public education and by introducing minimum energy performance standards for energy-intensive household appliances such as air conditioners and refrigerators.[63] As for the industries, the government plans to promote resource-efficient buildings through incentive schemes for property developers[64] and make energy labelling schemes mandatory.[65] Public-sector buildings will be required to achieve the highest green-mark ratings by 2020.[66]

In a nutshell, Singapore's approach towards climate change relies on the use of governance, economics and market forces. The private sector is being encouraged to take the lead in offering market-oriented solutions even as the state continues to create and improve awareness among households and the industries on the importance of energy efficiency. At the same time, the public sector is setting the standard when it comes to its energy consumption behaviour and through its

---

[61] Ibid: 125.

[62] Ibid: 26.

[63] MEWR; MND, 2009: "A Lively and Liveable Singapore: Strategies for Sustainable Growth", at: http://app.mewr.gov.sg/data/ImgCont/1292/sustainbleblueprint_forweb.pdf (20 January 2013): 41.

[64] Ibid.

[65] NEA, "About Mandatory Energy Labelling", at: http://app.nea.gov.sg/cms/htdocs/category_sub.asp?cid=258 (18 June 2011).

[66] MEWR; MND, "A Lively and Liveable Singapore: Strategies for Sustainable Growth", at: http://app.mewr.gov.sg/data/ImgCont/1292/sustainbleblueprint_forweb.pdf (20 January 2013).

procurement of facilities and equipments by adopting energy efficiency practices. Round things off, at the policy level, the government's inter-agency collaboration approach serves to bring key capabilities, knowledge and networks to the private sector.

5.3.1 Impediments, Rationale and the Way Forward

While Singapore has taken concrete steps to crystallise its climate change mitigation efforts, questions linger regarding their efficacy. For instance, how effective could economics and market forces be as long-term strategy for Singapore? Is it plausible to assume that as soon as or when an economic recession kicks in, household consumers for instance would resort to buying goods and consumer services that though not environmentally friendly are cheaper? Will the private sector, especially small and medium enterprises (SMEs), sustain its ecofriendly practices during periods of economic downturns?

It is possible that households with less financial means and smaller SMEs would cease to prioritise ecofriendly fiscal practices when the issue of their social and economic survival rather than the health of the environment becomes a matter of priority. As the security axiom goes, the threat is real only when it poses clear and present danger. Environmental security issues such as climate change are matters of security that many perceive as threat, but in the future and not in the short-to-medium term. This has been a perennial problem for governments—how do they convince the citizenry and the industries of the security threat posed by climate change?

This question therefore suggests that climate change policies enacted at the national level may prove to be ineffective when they lack the moral support of all stakeholders such as the citizenry and the industries. For instance, a country's and, in fact, the global economies are typically consumer oriented. This implies that an economy is sustainable only so long as consumers consume the goods and services produced by its manufacturers and offered by its retailers. The absence of such demand and supply would result in an economic meltdown. Unfortunately, the very nature of such an economy leaves a trail of environmental degradation. The health of the environment and therefore the climate is arguably a function of the state of the global economy—the higher the consumption, the more the resources (be it energy or natural minerals) needed to produce goods and services. This has been the case in the past. As Allenby rightly points out, "institutionally, we are beginning to recognise that the scale of human economic activity is for the first time fundamentally affecting a number of basic global and regional physical, chemical, and biological systems" (Allenby 2000: 5).

Singapore's national policies such as its energy efficiency programmes, recycling, green mark building, green labelling schemes and ecofriendly purchasing promote responsible consumption but may not leave a lasting impact on consumer behaviour. According to the traditional Rational Choice Theory,[67] consumers participate in an economic and financial transaction believing that they will reap maximum benefit from the utility of the purchased goods and services. As such, consumers would only be incentivised to purchase goods and services that offer them the highest value for their money. It is therefore safe to assume that consumers would not be moved to purchase goods such as ecofriendly vehicles or electronic household appliances should they cost more, and even if these goods and services were to leave a detrimental effect on the environment.

Quite to the contrary, Amartya Sen suggests that consumers may not only be driven by the pursuit of the maximum utility of goods and services (Sen 2008: 37). He argues that consumers also make decisions based on intrinsic values. Sen's argument is further reinforced by Herbert Simon's Theory of Bounded Rationality (Simon 1972). Simon argues that consumers or rather humans are irrational creatures and that they make decisions based on incomplete information about alternatives (Simon 1991: 163). The arguments put forward by Sen and Simon provide a note worthy of consideration for policy-makers and the security community, as their theses suggest that consumers might alter their behaviour if they could appreciate that the 'utility' of the goods and services that they have consumed can, in fact, be detrimental to themselves and their future generations.

Were Sen's and Simon's theses to be pursued by the policy-making community, they could alter the manner in which the community engages and markets their case for more sustainable ecofriendly practices to households and the industries. From the security community's perspective, the notion that the threat posed by climate change is real and has already affected the present generation as well as those to come would bring a sense of urgency to the public. There are already many reported cases of island-states in the Pacific that have either suffered or will be seriously threatened by rising sea water levels.[68] One area of concern is the vulnerability of fresh water reservoirs to the incursion of salt water. One such example is the Pacific nation of Tuvalu, whose agriculture has been severely affected by the incursion of salt water consequent to rising sea waters.[69] Closer to home, two non-governmental organisations (NGOs) in the Philippines—Green Coalition and Tik Tok For Climate Change Actions—claim that the shores along the coastlines of three Caraga provinces on the island of Mindanao have slowly

---

[67] Scott, John, "Rational Choice Theory", at: http://www.soc.iastate.edu/Sapp/soc401rationalchoice.pdf (30 June 2013).

[68] Chapman, Paul, "Enrire Nation of Kiribati to be Relocated Over Rising Sea Level Threat", in: *The Telegraph* (7 March 2012), at: http://www.telegraph.co.uk/news/worldnews/australiaandthepacific/kiribati/9127576/Entire-nation-of-Kiribati-to-be-relocated-over-rising-sea-level-threat.html (2 June 2013).

[69] Adams, Jonathan, "Rising Sea Levels Threaten Small Pacific Island Nations", in: *The New York Times* (3 May 2007), at: http://www.nytimes.com/2007/05/03/world/asia/03iht-pacific.2.5548184.html (2 June 2011).

been disappearing due to the rising seawater.[70] These incidents could serve as an indicator or measure of the gravity of the security threats posed by climate change. Under these circumstances, it can also be argued that security responses to climate change can no longer be limited to mitigation strategies alone. Matters have come to a head, so that states are now encouraged to channel their national resources towards focusing on adaptation strategies.

On this note, it can be argued that, for national climate change policies to be truly effective and sustainable, environmental citizenship has to be inculcated into the hearts and minds of not only Singapore citizens but also locally owned industries in the country. In other words, the state needs to convince all stakeholders of their civic responsibility towards the environment and to communities beyond its state borders.

5.3.1.1 Environmental Security

The security threat posed by a declining climate can no longer be considered as peripheral to larger traditional security issues. Since the end of the Cold War, the security literature has observed that states have been increasingly under the risks of non-traditional security threats such as energy security, human trafficking and smuggling, cross-border conflicts, pandemics, and cybersecurity. To the very same list, environmental security should also be added.

In the interest of setting the intellectual and policy foundation of this paper, and its basic conception, 'environmental security' refers to the "intersection of environmental and national security considerations at a national policy level" (Allenby 2000: 5). Others such as Foster articulate environmental security as "… the intellectual, operational, and policy space where environmental conditions and security concerns converge" (Foster 2005: 40). Renner goes further to elaborate that the focus of any policy research on environmental securtity should be contextualised against the core issues and concerns of the day (Renner 2006: 1–16).[71] In this regard, environmental security could mean many things to many people. For some, it has a direct impact on national security while others may prefer to analyse it through the lens of violent conflict, just as some others may prefer to study its impact on global security.[72] Notwithstanding its myriad definitions, the security literature relating to environmental security generally recognises the need to introduce it as an entity for defense and security policy considerations.

---

[70] "Some Coastal Areas, Islands in Caraga Sinking due to Rising Sea Water Level", in: *Manila Bulletin* (22 January 2011), at: http://www.mb.com.ph/articles/299986/some-coastal-areas-islands-caraga-sinking-due-rising-sea-water-level (8 May 2011).

[71] Renner, Michael, 2006: "Introduction to the Concepts of Environmental Security and Environmental Conflict", at: http://www.envirosecurity.org/ges/inventory/IESPP_I-C_Introduction.pdf (14 May 2011).

[72] Ibid.

For a culturally diverse and high-population density urban city-state such as Singapore,[73] it would perhaps be useful to study the impact of climate change on social cohesion.

5.3.2 *Ecological Citizenship and Green Political Thought: Strengthening Social Resilience*

The notion of ecological citizenship has its roots in the ideas of Andrew Dobson's seminal work (Dobson 2003: 61–68). In reviewing Dobson's concept of ecological citizenship, Bell explains that, in its most basic form, ecological citizenship is an extension of the liberal's notion of rights, which extends the concept that includes civil, political and economic rights to the realm of environmental concerns as well (Bell 2003: 2). In short, it refers to the individual's right to access environmental goods and to be protected from environmental bads. The relevance and significance of this concept is twofold. First, it extends the liberal's universal notion of basic human rights to include the environment and ecological realms. Second, and perhaps more importantly, it suggests that the right to access the environment can no longer be confined to one's private realm or within a country's state boundaries, and should instead be considered as a factor in one's political relationship with the environment. This also suggests that ecological citizenship binds the common human experience beyond physical state boundaries.

Other green theorists such as Neil Carter further argue that environmental citizenship is in fact the basis of a sustainable society (Carter 2007: 65). Carter observes that "… there is a consensus over the need for active ecological citizenship because of the recognition that the transition to a sustainable society requires more than institutional restructuring; it also needs a transformation in the beliefs, attitudes and behaviour of individuals."

Ecological citizenship then requires an informed citizenry that understands its roles and responsibilities in relation to its environment and ecology. As pointed out by Carolan, "… the value of this form of citizenship comes in its ability to lead to long-lasting behavioral change, versus the superficial change that emerges in response to dis/incentives (which immediately cease once the reward/punishment is revoked)".[74]

---

[73] Singapore is ranked second, behind China and Macau SAR, in terms of population density, according to the UN. See UN, "World Population Prospects: The 2010 Revision", at: http://esa.un.org/unpd/wpp/Sorting-Tables/tab-sorting_population.htm (5 July 2011).

[74] Carolan, Michael, "Ecological Citizenship and Tactile Space: The Epistemic Significance of the Lived Experience", at: http://www.michaelmbell.net/suscon-papers/carolan-paper.doc (21 June 2011).

5.3.2.1 Ecological Citizenship in Singapore

What Singapore needs is the kind of environmental citizenship that is rooted in its sociopolitical history and intellectual tradition. It would be dangerous for the nation if its citizenry were to adopt western liberal notions of citizenship in its entirety without first adapting it to their own sociopolitical imagination. Thus, for the effective adoption of values and rights that equips the citizenry with ecological citizenship, the notion of ecological citizenship has to be understood not as a different permutation of political and social citizenship but as imbued with the present notion of a country's political identity.

This, however, does not imply that Singapore lacks awareness of environmental degradation or even climate change. Since it acceded to the Kyoto Protocol in 2006, the government has worked closely through the different ministerial agencies and grassroot organisations to reach out to the citizenry in an effort to inculcate ecofriendly practices. Among other things, there have been numerous grassroots activities that promote practices such as recycling, energy efficiency practices and fuel-saving in the country. There is also some measure of civic participation in Singapore. Civil society organisations (CSOs) and NGOs such as ECO Singapore, Nature Society (Singapore) and the SEC have been actively engaging the citizenry in an effort to raise awareness about the environment and climate change. These organisations have also been receiving support from the government and other ministerial agencies.

It should be noted though that, while such efforts may be sufficient to raise civic awareness, they may not be enough to lead to a sustainable society. Such a society requires more than just civic awareness—it requires the socialisation of green values into the citizenry's political thinking and identity. In other words, to be a Singaporean would also have to mean being ecofriendly.

5.4 Towards a Robust Sustainable Society: Some Policy Recommendations

The UNFCCC is one of many examples that shows that there is global concensus on the recognition of the need to adopt a global strategy to mitigate and adapt to climate change. In his speech at the *Singapore International Energy Week 2010*, Prime Minister Lee Hsien Loong stressed the need to apply economic principles and market-based approaches to encourage the private and public sectors as well as households to adopt ecofriendly practices.[75] The Prime Minister's speech demonstrates the state's political will to ensure that its energy security policies are

---

[75] "Transcript of Prime Minister Lee Hsien Loong's Address at the Singapore Energy Lecture, Singapore International Energy Week, 1 Nov 2010", at: http://2010.siew.sg/sites/singapore.iew.com.sg/files/PMLeeSpeechatSEL.pdf (14 April 2010).

ecofriendly and sustainable, at least at the level of state–state, state–industry and state–society relations. Although Singapore has published a sustainable development blueprint, the authors propose the following recommendations to help develop a more robust sustainable society.

5.4.1 Education Towards a Green Lifestyle

It would be but appropriate that the first steps towards the creation of a sustainable society should begin in the country's schools. Climate change and environmental science should be included as core subjects in the national curriculum in public schools, and students should be taught about the implications of their social and economic behaviour on the environment—it is only through schools that a future Singaporean society can be socialised towards a sustainable ecofriendly society. To this end, and to emphasise their importance and relevance, the subjects of climate change and environmental science could be given the same weightage as core subjects such as English and Mathematics.

5.4.2 Direct Participation in Environmental Decision-Making

Accountability is an important feature of a sustainable society. It is only when members of the society hold themselves, the state and the international community accountable for the impact of their socioeconomic actions on the environment that meaningful change can be achieved in the long run. To this extent, the government and the bureaucracy should encourage greater civic participation in the environmental decision-making process.

The Singapore state has a good record of civic engagement with members of the public. In fact, it has often reached out to the public, channeling consultation through advisory committees and other consultative forums. However, key decisions, particularly those of national interest, have always been made by the legislature and in consultation with members of the public. As a sustainable society would likely require greater public involvement in this area, members of the public could be roped in, to begin with, at the grassroots consultation forums or even at the level of community development councils for schemes such as neighbourhood infrastructure and design planning.

5.4.3 Holding Businesses Accountable

In Singapore, the industrial and business sectors are responsible for the majority consumption of energy and electricity sources, with households accounting for only a small percentage of national consumption. The state could therefore encourage greater civic participation by encouraging more CSOs and NGOs to act as government watchdogs and whistleblowers when businesses are found to be environmentally irresponsible.

5.4.4 Access to Justice in Environmental Matters

In furtherance of the point above, the state could formalise legal avenues and recourse, where members of the public and civil society are able to pursue legal actions against business corporations or other private organisations that are found to be guilty of non-environmentally friendly actions.

5.5 Conclusion

Singapore's climate change strategy is largely premised on economics and market-based solutions, and it appears to be comprehensive because it is fully supported by a 'whole-of-government' approach. The state has clearly demonstrated its political will to ensure that the nation is prepared to do its bit in the global effort towards mitigating climate change. However, the question remains whether the state's holistic approach would be sustainable if the same rigour and political will are not demonstrated at the grassroots level. After all, in the larger scheme of things, households and businesses, as consumers and producers of goods and services, are largely responsible for the consumption of energy and environmental resources.

This chapter argues that any attempts to mitigate climate change can only be meaningfully achieved in a sustainable society—one whose sociopolitical identity has been inculcated with green values. A sustainable society, in short, is based on ecological citizenship that is deeply rooted in the belief that every human being has a political relationship to his/her environment.

To achieve such a society, the paper recommends the institutionalisation of climate change in public schools through the national curriculum, involvement of members of the public in the nation's environmental decision-making process, empowering CSOs and NGOs as government watchdogs and whistleblowers in case the public and private sectors do not adopt ecofriendly practices and, finally, provision of a legal framework for the prosecution of private sector organisations that violate environmental laws within and outside Singapore.

References

Allenby, Braden, 2000: "Environmental Security: Concept and Measurement", in: *International Political Science Review*, 21,1: 5–21.

Bell, Derek R., 2003: "Environmental Citizenship and the Political", Paper presented to the Economic and Social Research Council (ESRC) Seminar Series on 'Citizenship and the Environment', Newcastle, UK, 27 October, at: www.ncl.ac.uk/environmentalcitizenship/papers/Bell.doc (4 July 2013): 2.

Carter, Neil, 2007: *The Politics of the Environment: Ideas, Activism, Policy*, 2nd ed. (Cambridge, UK: Cambridge University Press): 65.

Dobson, Andrew, 2003: *Citizenship and the Environment* (Oxford, UK: Oxford University Press).

Foster, Gregory D., 2005: "A New Security Paradigm", in: *WorldWatch* (January/February), 36–46.

Renner, Michael, 2006: "Introduction to the Concepts of Environmental Security and Environmental Conflict", in: Kingham, Ronald A. (Ed.): *Inventory of Environment and Security Policies and Practices* (The Hague: Institute for Environmental Security): I-C 1–16.

Sen, Amartya, 2008: "Rational Behaviour", in: Durlauf, Steven N.; Blume, Lawrence E. (Eds.): *The New Palgrave Dictionary of Economics*, 2nd ed. (New York: Palgrave Macmillan).

Simon, Herbert A., 1972: "Theories of Bounded Rationality", in: McGuire, C.B.; Radner, Roy (Eds.): *Decision and Organization* (Amsterdam: North-Holland Publishing Company): 161–176.

Simon, Herbert, 1991: "Bounded Rationality and Organizational Learning", in: *Organization Science*, 2, 1 (February): 163–173.

Abbreviations

A*STAR	Agency for Science, Technology and Research
BCA	Building and Construction Authority
BH	Berita Harian
CCAP	Climate Change Awareness Programme
CO_2	Carbon dioxide
Cleantech	Clean technology
CSO	Civil society organisation
E^2PO	Energy Efficiency Singapore Programme Office
EDB	Economic Development Board
EENP	Energy Efficiency National Partnership
EIA	U.S. Energy Information Administration
EMA	Energy Market Authority
ERI@N	Energy Research Institute at Nanyang Technological University
ESI	Energy Studies Institute
ESRC	Economic and Social Research Council
Gt	Gigawatt
HDB	Housing and Development Board
IDA	Infocomm Authority of Singapore
IEA	International Energy Agency
IMCCC	Inter-Ministerial Committee on Climate Change
JTC	Jurong Town Corporation

LNG	Liquefied natural gas
LTA	Land Transport Authority
MEWR	Ministry of the Environment and Water Resources
NCCS	National Climate Change Secretariat
NEA	National Environment Agency
NGO	Non-governmental organisation
NTS	Non-traditional security
NRF	National Research Foundation
NTU	Nanyang Technological University
NUS	National University of Singapore
PUB	Public Utilities Board
QatarGas	QatarGas Operating Company Limited
R&D	Research and development
RIMA	Centre for Research on Islamic and Malay Affairs
RSIS	S. Rajaratnam School of International Studies
SCEM	Singapore Certified Energy Managers
SEC	Singapore Environment Council
SEI	Singapore Environment Institute
SERIS	Solar Energy Research Institute of Singapore
SLNG	Singapore LNG Corporation
SMEs	Small and medium enterprises
Tcm	Trillion cubic metre
UN	United Nations
UNFCCC	United Nations Framework Convention on Climate Change
URA	Urban Redevelopment Authority

Chapter 6
Empowering the People: Towards the Inclusion of a Global Civil Society in a New Climate Change Regime

Eulalia Han

Abstract The United Nations Framework Convention on Climate Change (UNFCCC) acknowledges the global nature of climate change and encourages the widest participation in tackling this issue. However, climate change initiatives are still targeted at the national level, and the disconnect between national and global initiatives have resulted in huge disparities between the speed and effectiveness of policies among signatory countries. This chapter supports the development of a global civil society to encourage a concerted effort between states. More importantly, it recognises that citizens remain at the core of any such effort and calls for the inclusion of 'green values' as part of the nation-building exercise.

Keywords Climate change · Environmental rights · Global civil society · Green values · Mekong River · 'Tragedy of the commons' · UNFCCC

6.1 Introduction

The United Nations Framework Convention on Climate Change (UNFCCC)[1] is an international treaty where signatory countries agree to cooperate on combating and coping with the impacts of climate change. Developing sustainable practices now occupies national strategies as countries look to reduce their carbon footprint. To date, 195 countries are Parties to the Convention, "… acknowledging that the global nature of climate change calls for the widest possible cooperation by all

---

[1] UNFCCC, "Cooperation & Support", at: http://unfccc.int/cooperation_and_support/items/2664.php (11 July 2013).

E. Han (✉)
Energy Studies Institute (ESI), National University of Singapore (NUS), 29 Heng Mui Keng Terrace, Block A #10-01, Singapore 119620, Singapore
e-mail: esihne@nus.edu.sg
URL: http://www.esi.nus.edu.sg/about-us/our-researchers/dr-eulalia-han

N. A. Putra and E. Han (eds.), *Governments' Responses to Climate Change:* *Selected Examples From Asia Pacific* 10, SpringerBriefs in Environment, Security, Development and Peace, DOI: 10.1007/978-981-4451-12-3_6, © The Author(s) 2014

countries and their participation in an effective and appropriate international response".[2] To achieve this, the UNFCCC seeks to promote cooperation and exchange of information related to climate change, invest in public awareness and "encourage[s] the widest participation in this process, including that of non-governmental organi[s]ations".[3]

The nation-state, while a contested term, remains the only legal entity in the international community (Anderson 1983; Giddens 1984; Habermas 1989; Walby 2003). This means that regardless of its political system, a state is the only recognised representation of its citizens in international relations or state-to-state negotiations. Global efforts to combat climate change have therefore focused on state-led initiatives and states "continue to dominate the procedures and the substance of interaction on key sovereignty-related issues" in United Nations' (UN) negotiations (Clark et al. 1998: 6). Non-governmental organisations (NGOs) or civil society organisations (CSOs), while gaining importance and presence in international negotiations, still remain at the periphery in the formal policy and implementation process.

There is, however, growing recognition of the importance of engaging civil society when addressing non-traditional security (NTS) threats such as energy security and climate change (Baker 2006; Caballero-Anthony et al. 2012). Climate change, as an NTS threat, is a result of "human-induced changes" from energy use (Karl/Trenberth 2003: 1719). Therefore, the development of an ecological society, or the idea of ecological citizenship, requires the participation of citizens so as to reconcile environmental protection, economic activities and social progress (Dobson 2007; Baker 2006).

6.2 Global Civil Society as a Way Forward

Many of the current climate change efforts are targeted at the private sector. As much as big businesses contribute mostly to rising carbon dioxide levels, a strict focus on the role of the private sector in addressing climate change overlooks the reality that production is also driven by consumption. This approach disempowers citizens as it potentially gives the impression that the main responsibility of addressing climate change is that of governments and corporations. Moreover, the UNFCCC has stipulated the importance of education, training and public awareness so as to promote public participation in addressing climate change at the national, regional and international levels.[4] As signatories to the Convention, states

---

[2] UNFCCC, "Report of the Conference of the Parties on its Seventeenth Session, Held in Durban from 28 November to 11 December 2011", FCCC/CP/2011/9/Add.1 (15 March 2012), at: http://unfccc.int/resource/docs/2011/cop17/eng/09a01.pdf (21 January 2013): 2.

[3] UN, "United Nations Framework Convention on Climate Change" (1992), FCCC/INFOR-MAL/84, at: http://unfccc.int/resource/docs/convkp/conveng.pdf (21 January 2013): 6.

[4] Ibid: 10.

should equally involve both the private and public sectors in the nation's climate change strategies.

The previous chapters also note inconsistencies and disparities in government policies regarding the issue. The development of sustainable communities is present in national strategies to address climate change. However, national targets and the process of achieving that goal remain unclear. Within the international community, definitions of sustainable development remain broad, with emphasis on different aspects of sustainability. For instance, the Brundtland Commission defines sustainable development as "development which meets the needs of the present without compromising the ability of future generations to meet their own needs" (WCED 1987: 43). At the crux of this definition lies the issue of inter-generational equity in that humanity's present choices should not negatively affect future generations (Berke/Conroy 2000; Bradshaw/Winn 2000). Meanwhile, the Institute for Sustainable Communities (ISC) in Washington defines a sustainable community as "one that is economically, environmentally, and socially healthy and resilient".[5] The Organisation for Economic Co-operation and Development (OECD) also provides a comprehensive and adequate working definition of sustainable development "as a significant change in how people and governments perceive their activities, their roles and responsibilities: from primary emphasis on increasing material wealth to a more complex, interconnected model of the human development process" (Strange/Bayley 2008: 30). On the other hand, in the context of Singapore, sustainable development "means achieving both a more dynamic economy and a better quality living environment, for Singapore's now and in the future" (MEWR/MND 2009: 12).

To effectively address climate change, however, government strategies should be clear, coherent, well defined and reflective of local concerns. If there is consensus that the scale of human economic activity results in significant changes in our environment, and that sustainable practices should be institutionalised and acknowledged by citizens, then CSOs should be recognised as playing a potentially important role in this process. A well-developed civil society could enhance political responsiveness to local issues and, in turn, the government could formulate green growth policies that are prescriptive rather than descriptive (Diamond et al. 1997; Clark et al. 1998). In addition, as climate change is a global issue, a concerted effort to tackle environmental issues will require discussions that cut across states. For this to be successful, the actors involved in these exchanges must be seen to reflect domestic concerns and, more importantly, be able to protect interstate relations, especially regarding the laws of sovereignty and in some regions such as Southeast Asia, where the policy is of non-intervention and non-interference. A global civil society (GCS) would be a viable representation as it is "the sphere of ideas, values, institutions, organisations, networks, and individuals

---

[5] ISC, "What is a Sustainable Community?" (2012), at: http://www.iscvt.org/what_we_do/sustainable_community/ (6 July 2013).

located *between* the family, the state, and the market and operating *beyond* the confines of national societies, polities, and economies" (Anheier et al. 2001: 17).

6.3 The Role of Global Civil Society in the New Climate Change Regime

GCS organisations have emerged as a powerful influencing force in international politics. Seen as a third alternative to the organisation of international relations, GCS organisations "question the monopoly of the nation state over the lives of its people", "challenge the workings of international institutions" and "act as the guardians of a morally informed consensus on the minimum that is due to human beings" (Chandhoke 2001: 41). This is not to assume, however, that these organisations are always autonomous from the institutions they seek to challenge or that they can always provide a viable alternative to the existing world order. Afterall, GCS might still be confined by the same relations that characterise international politics. Moreover, there are concerns that the practices of GCS could potentially deny states of their rights to sovereignty, or challenge the values accorded to states and the system of states (Lipschutz 1992: 391; Falk et al. 1993; Chandhoke 2001: 36).

Concerns have been raised, particularly in developing countries or by Third World governments, that the values espoused by GCS reflect a narrow elite from influential states, and the institutionalisation of these values is being seen as an 'imposition' on the former (Chandhoke 2001: 51–52). Another supposed criterion for vibrant civil societies is that they "require strong and stable states as a pre-condition to their very existence" and that "states constitute the limits of civil society .... [as only] they can provide the conditions within which the civil society agenda is realised" (Chandhoke 2001: 51).

To counter the limits of GCS, we have to arrive at a common understanding of how we should contextualise and locate GCS within the international system. If GCS is posed to provide an alternative to states and institutions, it has to work within the confines of established rules of international law, especially when it concerns issues such as addressing climate change that require cooperation between all existing stakeholders be it the states or the UN. The most appropriate way to understand the role of GCS then is that it is "a state-centric system of international relations", filling the gap between state and society while crossing state boundaries (Chandhoke 2001: 52). It should also be understood as "an alternative organising principle for world politics, based on new constitutive rules and institutional forms" (Conca/Lipschutz 1993: 9). GCS could be made up of a conglomerate of local CSOs or exist as an entity on its own, negotiating between two or more states. This strategy is significant for addressing global issues such as climate change, as it provides a platform for climate change strategies, established at the local-national level, to be discussed at the regional or international level where many countries are dealing with the same issues. This is important because climate change is a global issue and requires a concerted effort to tackle. GCS could additionally target specific issues that individual states face and specifically cooperate with states facing similar problems. This is especially significant given the notion of "inadequate state" in handling climate change unaided, which questions the traditional notion of sovereignty and whether the latter can be maintained within an ecological frame (PRIO 1989; Elliott 2001: 143). The solution to this could reside in "the concept of cooperative sovereignty" in areas where it is inadequate to exercise sovereignty alone (Elliott 2001: 143).

6.4 Empowering the People: The First Step to a Sustainable Global Civil Society

For the development of a sustainable GSC as a platform for countries to effectively cooperate on addressing climate change, citizens need to be aware of and appreciate their position as members of their local and international community. Empowerment is now increasingly seen as an important element in equitable climate change strategies and to reduce vulnerability experienced by directly affected communities (World Bank 2000; Skoufias 2003: 1100; Tompkins/Adger 2004; Thomas/Twyman 2005: 118–120). 'Green values' therefore should form part of the nation-building exercise and not simply be an addition to current national frameworks.

There are two strategies to encourage effective participation from the people. First, discussions on the role of developed and developing countries in addressing climate change should be included in the narratives of either formalised education institutions or local communities. In addition, issues such as equity and justice should be considered when thinking about mitigation and adaptation strategies (Adger 2000: 754–756; Thomas/Twyman 2005: 118–120). Second, GCS can then facilitate crossboundary conversations and cooperation by encouraging the participation of local representatives. It has been noted that the predominant focus of the UNFCCC has been at the "national and larger scales [which could be] potentially [leaving] a vacuum at sub-national levels with regard to the equitable nature of the impacts of adaptive strategies to climatic change" (Thomas/Twyman 2005: 115). Local communities' representation would not only effectively fill that void but also, when represented at the level of GCS, allow for these issues to be addressed at a larger scale (Mortimore/Adams 2001: 55).

6.4.1 Whose Right?

Before discussing the role of developed and developing countries in addressing climate change, it is important to mention the debates surrounding the right to exploit nature's reserves and the reality of overextraction. These issues are important to understand the factors that have influenced the divide between the way developed and developing countries approach climate change.

Economic growth in developed countries has been achieved at the expense of the environment. The industrial revolution experienced during the nineteenth century and again in the 1970s made way for technological, regulatory and economic forces to enter our presence. The desire for material progress saw the conversion of forested areas for agricultural production and the use of natural resources to spur the increase in energy demands (Baker 2006: 12–26). With this, the concept of 'property' was further engrained in our consciousness, implying the "rights of *exclusion*", the "rights of *use*" and the "rights of *disposition*" (Brown 2004: 13). This sense of property, ownership, belonging and right to inherit the earth's riches precede modern times and can even be traced to the propagation of Abrahamic religions.

British philosopher John Locke, for example, sees the Bible as authority and based his ecological philosophies on human's natural right to extract and cultivate the land's resources (Laslett 1988). He views the cultivation of resources as an obligation to preserve mankind, as it was much more productive than land in its natural state (Brown 2004: 13). These views were also shared by Aristotle and Descartes, who believed that the fundamental difference between man and his environment was the former's ability to think, and hence that his surroundings were "devoid of reason" (Descartes 1927: 360). The rise of philosophical utilitarianism saw philosophers such as Jeremy Bentham and John Stuart Mill providing a somewhat more complex way of looking at man's relations with the Earth. Their argument that all "living things that are capable of suffering should be placed at the centre of moral judgement", coupled with their philosophy that "property was to produce the greatest number of happiness", is still narrowly focused on an all-human perspective (Brown 2004: 14).

These ideas have informed much of the material, industrial and economic progress that we have experienced for a few centuries. Some traditional societies, on the other hand, are founded on 'nature religions' or pantheism. These societies, which might be highly dependent on the environment for their livelihood, therefore base their philosophy on living in oneness with nature. Even in capitalist economies, discussions over models of resource justice, unequal ecological exchange and securing livelihood rights are increasingly tilting to accommodate the climate change agenda (Sachs/Santarius 2007). The original 'right' to extraction, production and consumption of the Earth's resources—now questioned as overextraction, overproduction and overconsumption—has inevitably led to humanity experiencing the effects of an environment that is unable to replenish its resources in line with human actions. Overexploitation has now led to what (Hardin 1968) had termed as "the tragedy of the commons".

6.4.2 Hardin and 'The Tragedy of the Commons'

'The tragedy of the commons' is a model that explains the perils of overexploitation and the potential outcome of the complete degradation of the commons. Although his focus was on overpopulation, Hardin highlights the point that "'rational' individual actions can lead to 'irrational' collective practices resulting in catastrophic over-exploitation of common resources" (Greene 2005: 459). Speaking on overpopulation, he puts forth a scenario whereby overpopulation means that the current population will have to give up some of the 'good' that they have received and rely on technology to cater to the extra needs of the population that are unsustainable. Essentially, he argues that the "population problem has no technical solution; it requires a fundamental extension in morality" (Hardin 1968: 1243).

Hardin's argument stems from the idea that unregulated access to the commons such as lakes and pastures allows for individuals to exploit resources at the maximum level for their own benefit. As a result, only these individuals benefit while the entire community has to pay the price of overexploitation. The 'tragedy of the commons' "is that this depletion of 'open access' common resources can continue remorselessly to its destructive conclusion, even if each user involved is well intentioned, well informed, and exercising only its traditional and legal rights. Unilateral acts of public-spirited restraint are insufficient to tackle the problem. If the rest of the user community continues in its old ways, the public-spirited suffer along with the selfish without even having benefited from the 'good times' in the meantime" (Greene 2005: 459).

6.4.3 The Role of Developed and Developing Countries in Addressing Climate Change

Recognising arguments surrounding the rights to extraction, production and consumption alongside Hardin's 'tragedy of the commons' allow us to understand the factors that have contributed to the current contentions between developed and developing countries in addressing climate change. Economic globalisation, for example, has exacerbated vulnerabilities to climate change as it drives industrial production, increases energy consumption and, arguably, locks economies that are not able to 'catch up' to these advancements (Rostow 1990; O'Brien/Leichenko 2000: 227). Three major economic regions, namely Northeast Asia (initially led by Japan), the European Union and North America (the United States, Canada and Mexico), account for more than two-thirds of the world's trade (O'Brien/Leichenko 2000: 226). With increased economic activity among these countries, regions such as South Asia and Sub-Saharan Africa have been marginalised from the global market, especially in terms of foreign direct investments.[6] However, although the latter regions contribute the least to carbon dioxide emissions,[7] as they are one of the lowest users of energy,[8] the Sub-Saharan and South Asia regions experience the brunt of climate change (McGuigan et al. 2002).

There is widely accepted consensus that poorer developing countries are most at risk and vulnerable to the effects of climate change (IPCC 2001; Richards 2003; Thomas/Twyman 2005: 116–120). In terms of the four large areas covered in the climate change debate (mitigation, adaptation, adaptive capacity and vulnerability), developing countries are viewed as being most susceptible to the economic ramifications of climate change and having the least technical capabilities to deal with the latter's impact. While economic and technological developments have greatly aided the modernisation process and benefited a significant portion of humanity, developed countries should also claim responsibility for the damages to the environment that modernisation has caused. Increasingly, questions regarding the morality of extraction, consumption and (re)distribution of natural resources have also been brought to the fore. The arguments of cosmopolitan justice, as espoused by the assumptions of 'universal human rights' by Thomas Pogge and Immanuel Kant, underlie these suppositions and, in this instance particularly, the right to exploit existing resources at the expense of another who is equally entitled (Gregor 1996; Pogge 2000).

The issues discussed so far have given rise to discontent between developed and developing countries at the UNFCCC and other international negotiation forums. There is a stark disparity, especially on the issue of equity in mitigating climate change and the concept of environment security. Developed countries view equity in terms of developing countries' 'participation' in mitigation strategies and managing global carbon trade (Muller 2002). Developing countries, on the other hand, "favours a per capita emissions rights based solution" and raise issues such as "'the right to emit' to reach a level of economic development which satisfies basic human rights" (Richards 2003: 4). The arguments posed by developing countries have greatly weakened their negotiating position at the UNFCCC, as their responses are usually reactive and defensive rather than proactive (Richards 2001: vii–viii).

Regarding environment security, developed countries tend to see environmental challenges as contributing to political conflict, and hence park it under the security agenda of the military (Soroos 1997; Elliott 2001: 147). A militarised environment security agenda makes very little room for understanding the complexity of how poverty, environmental degradation, injustice and conflict interacts (WCED 1987; Elliott 2001: 147). Even when developed countries recognise the importance of

---

[6] World Bank, "Foreign Direct Investment, Net Inflows (BoP, Current US$)" (2012), at: http://data.worldbank.org/indicator/BX.KLT.DINV.CD.WD/countries?display=default (25 January 2013).

[7] World Bank, "Climate Change" (2013), at: http://data.worldbank.org/topic/climate-change?display=default (25 January 2013).

[8] World Bank, "Energy & Mining" (2013), at: http://data.worldbank.org/topic/energy-and-mining?display=default (25 January 2013).

investing in renewable energy and energy efficiency, the philosophy that drives that push is only partly aiming to address climate change; the thrust is mainly to reduce dependency on energy imports and to be secure from volatile commodity prices. On the other hand, developing countries, especially those whose economy is based on their agricultural sector, greatly depend on climatic conditions for their livelihood. Therefore, while interstate and regional conflict persists, developing countries are more focused on mitigating the causes of environmental degradation rather than focusing on the conflict that might result because of climate change. They are also more concerned with poverty as both an influence and consequence of erratic climatic conditions, the question of justice and morality when it comes to the overexploitation of nature's resources, and improving their economy and electrification rate amid other predicaments.

The empowerment of citizens requires these narratives to continually feed into the nation-building agenda. From the subject of rights to inheritance to the problem of overexploitation and the impact of economic globalisation in accentuating the developed-developing countries divide, these discussions bring to light the interconnectedness of the international system, which when put together allows citizens to understand the causes and consequences of climate change from its many angles. From such a perspective, what would be the next step forward for people's participation in the larger climate change project? Individual efforts by states have proved futile at effectively addressing climate change on a larger scale. As noted, the weak negotiating capacity of developing countries has also undermined their bargaining power in holding the more influential and industrialised states accountable at United Nations forums. A more proactive negotiation strategy would require strengthening the negotiating capacity of developing countries through the provision of more negotiating resources in terms of institutional support, communication skills, information and research, so that they are better able to counter the asymmetry in current international discussions and compete with the more equipped and experienced developed countries (Richards 2003). As intervention is necessary to "enhance adaptive capacity or the ability to adapt to new or changing conditions without becoming more vulnerable", GCS could play an important role in supporting cooperation between communities from both developed and developing countries (Adger et al. 2003: 190).

6.5 Empowered Communities as Participants in Global Civil Society

Communities in developing countries who are most susceptible to the impact of climate change are not only passive victims. Reinforcing such ideas is potentially disempowering, as it creates "a discourse of vulnerability that undervalues and undermines the potential and actions of people facing significant disturbances by climate change impacts" (Thomas/Twyman 2005: 116). On the flip side, developed communities do practise self-restraint and do not always seek to maximise profits if they are able to appreciate the value and utility of their consumption (Sen 1999).

Research suggests that the extent to which communities are able to effectively address climate change is dependent on when they are given the 'tools' to adapt and respond to it (Tompkins/Adger 2004; Thomas/Twyman 2005). These 'tools' or 'head room' are present in the form of empowerment that, when people are aware of their habits of consumption and production and how that fits into the larger international community, will incentivise them more to reconceptualise their behaviour and cooperate with affected parties to arrive at an acceptable outcome. The presence of GCS can facilitate this process of empowerment and cooperation. The chapter's next section will present a short case study on the development of hydropower in Southeast Asia and how GCS could assist in the development of a common effective framework for the management of the Mekong River.

6.5.1 Development of Hydropower in Southeast Asia

As part of Southeast Asia's efforts to diversify its energy mix to counter volatile commodity prices, countries with huge hydropower potential have embarked on the development of hydro energy to secure their energy and financial needs. Cambodia, Myanmar and Vietnam are in the process of building dams along the Lower Mekong River, and China, which controls the Upper Mekong Basin, has also begun dam construction.[9] The Mekong River is able to provide the riparian states with revenue and alternative sources of energy. Laos' and Myanmar's future sale of hydro electricity to Thailand, for example, will be able to improve the countrys' social development programmes while Cambodia will be able to use additional hydropower to increase its rural electrification rate. The riparian states of the Lower Mekong Basin remain one of the poorest regions in the world, with 85 percent of its population living in rural areas and only half of the households having access to safe drinking water (MRC 2013: iii).

While there are many benefits to the development of hydropower plants along the Mekong River, there are also significant potential problems that could arise from their construction. These hydropower projects could have an adverse impact on the environment, as they could affect the survival of fish species and cause the destruction of fisheries, thus leading to a loss of livelihood and food sources (Baker 2012). Fish remains a staple source of protein for the millions of people in these riparian states, and the impact that these dams will have on fish migration is expected to pose a significant problem for more than ten million people (Baker 2012: 8). This would then see the eviction of people from their homes and encourage mass migration in affected areas. Large projects such as the Lancang

---

[9] For more information and a map of existing and proposed sites of dam construction on the Mekong River, see Save the Mekong, "Dams Locations and Status" (8 March 2009), at: http://www.savethemekong.org/issue_detail.php?sid=21 (25 January 2013).

dam in the Upper Mekong Basin have caused irregular flows downstream and hindered the siltation process (IRN 2002; Baker 2012). In Laos and Thailand, planted crops in the sedimentary soil are regularly washed away by the sudden release of dam water in the Upper Basin (Baker 2012: 5). This has also caused bank erosion. To date, the two most controversial dam projects, both in Laos, are the Don Sahong and the Xayaburi dams (Baker 2012: 5). These two dams have the potential to affect sediment capture and cause other ecological problems that cannot be confirmed.

For years, hydro energy was classified as renewable energy by some member states of the Association of Southeast Asia Nations (ASEAN) and as a renewable energy resource by others depending on installed capacity (Suryadi 2012: 2–5). This difference in classification would have implied a different understanding of what constitutes renewable energy and therefore accorded varying legal criteria to hydropower during the initial development stages of some of these dams. It was only during the 28th meeting of the Heads of ASEAN Power Utilities/Authorities (HAPUA) on 6 June 2012 that hydropower was officially regarded as renewable energy (Suryadi 2012: 4–5). Also, while Laos, Cambodia, Thailand and Vietnam have signed on to the Agreement on the Cooperation for the Sustainable Development of the Mekong River Basin in 1995, China remains a non-signatory.

6.5.1.1 The Mekong River Commission: Limitations

The Mekong River Commission (MRC) was established by Cambodia, Laos, Thailand and Vietnam to "cooperate in all fields of sustainable development, utilisation, management and conservation of the water and related resources of the Mekong River Basin" (MRC 2013: ii). China and Myanmar are currently dialogue partners with the MRC. In meeting the challenges ahead, the Commission contends that "the greatest threat to the environement is sustained poverty and associated marginalisation and exclusions, together with the conflict that likely will follow" (MRC 2013: 11).

The MRC involves national and international stakeholders in discussing pertinent issues surrounding the Lower Mekong River Basin such as agriculture, agricultural water management, analysing the multifunctionality of paddy fields, the impact of climate change on the Mekong Basin, environmental health and fisheries. While the MRC remains committeed to managing existing and potential problems of the Mekong River, the limited participation of selected national and international representatives proves inadequate when it comes to resolving certain issues. To begin with, the MRC hardly publishes or engages in fieldwork regarding the impact of dam construction on the livelihoods of the citizens in the riparian states or even considers the relationship between dam construction and rising flood levels. International Rivers, an international NGO, for instance, has cited many examples of how dam construction along the Mekong River's mainstream has threatened the river's ecology and the well-being of millions of people who are dependent on the river for income, food and transportation.[10]

Because the MRC was formed by the governments of Cambodia, Laos, Thailand and Vietnam, who in the first place initiated the development of hydropower in their respective countries, the role of the MRC is not to question the construction of the dams itself but to promote the sustainability of that development. Moreover, as ASEAN member states practise a policy of non-interference and non-intervention in another state's domestic affairs, the MRC might not have the authority to discuss and cooperate openly on issues such as the impact of dams on affected communities.

6.5.1.2 Role of Global Civil Society in the Development of the Mekong River

Although the threats to the livelihoods of people and the environment from dam construction are real, the local population that will be the most affected by such adverse impacts is excluded from the decision-making processes in the case of the Mekong River. GCS could prove effective in bridging the gaps between society and the state and between states, especially in Southeast Asia where the policy of non-interference is upheld. There is much research to suggest that "at the national and international levels, policy responses to climate change should be oriented towards creating or facilitating the emergence of 'head room' thus enabling, rather than inhibiting, local and regional level adaptation options" (Thomas/Twyman 2005: 121). GCS could provide this head room and facilitate crossboundary conversations and cooperation by encouraging the participation of local communities. So, how can GCS assist affected communities in the Mekong River riparian states?

The MRC (2010: 43) has identified that millions of people living along the Lower Mekong Basin "lack the opportunity to exercise their civil rights or access information. The mechanisms do not exist to allow them to take part in decisions on issues concerning water resource management, impacts of development or access to common goods such as lands and flooded forests". GCS would provide citizens with adequate tools and platforms for negotiations. Providing local communities with up-to-date research on the impact of the construction of dams on their livelihood and supporting a sustainable training programme that teaches them the skills of negotiation will strengthen their negotiating capacity. This will be first done at the local level, where local communities can openly discuss how hydropower development has the potential to lock them in their current poverty cycle and how remaining in poverty might force them to further exploit the river's resources for basic sustenance. Social resilience, after all, is key for any community to respond to environmental issues or to adapt to climate change (Tompkins/Adger 2004; Thomas/Twyman 2005).

---

[10] International Rivers, "Mekong Mainstream Dams" (2007), at: http://www.internationalrivers.org/campaigns/mekong-mainstream-dams (26 January 2013).

GCS can also facilitate communication between local groups of different states by playing the role of a mediator. For example, a group of representatives from affected communities in Laos and Thailand may be able to arrive at common solutions for solving the siltation problem created by the construction of dams. If successful, GCS can further mediate communication between local communities from Laos, Myanmar and China, for example, to derive sustainable management and development practices that can be agreed upon. The problem of siltation in the Lower Mekong River caused by current upstream activities will eventually affect water and soil conditions in Myanmar and China, and therefore cooperation between affected communities in these countries would be necessary. The difference between these initiatives, as compared to formalised negotiation between governments, is that GCS would ensure the inclusion of local communities in the decision-making process. Empowering local communities is important because it accords power and responsibility to the largest portion of the population and, in the case of the Mekong River, to communities who are the most affected by the development of hydropower in the region.

6.6 Greening Nation-Building, Creating Adaptive Policies

The discussions in this chapter reinforce the importance of empowered communities and the role of GCS in addressing global environmental issues. It suggests that green values should be a part of the nation-building exercise through discussions on the question of rights to extraction, production and consumption, the problems of overexploitation, and the role of developed and developing countries in the climate change debate as a whole. These discussions could be either institutionalised in formal education or embedded within local communities. The hope is that, through the greening of national institutions and conversations, governments will be inclined to create adaptive policies. It has been noted that to effectively address climate change,

> [p]ublic policies have an important role to play in fostering this ability. But for policies to be effective and to help people, the policies themselves must also give careful consideration to complex interactions and be able to adapt to conditions that can and cannot be anticipated. A policy that is unable to continue to perform in a dynamic and uncertain setting, or unable to detect when it is no longer relevant, is a policy that is more likely to hinder the freedom and capability of people to adapt to change (Venema/Drexhage 2009: 2).

Adaptive policies are policies that have the ability "to navigate towards successful outcomes in settings that cannot be anticipated in advance" and "anticipate the array of conditions that lie ahead through robust up-front design" (Swanson et al. 2009: 15). If governments are to be motivated by sustainability, committed to the UNFCCC and aspiring for a future where the impacts of our actions on humanity are viewed holistically (from the social, economic and environmental perspectives), what would be most important would be how people understand their place, interact with one another and adapt to change (Venema/Drexhage 2009).

The case study of the Mekong River highlights the potential role that GCS could play in addressing the problems associated with the development of hydropower plants. GCS can play a mediating role between local communities in different states without infringing upon the laws of sovereignty, non-intervention and non-interference. More importantly, unlike the MRC, GCS would enable open interaction with affected communities and a confrontation of deeper underlying issues such as the questions of equality, equity and entrenched poverty as a result of the building of hydropower dams along the Lower Mekong Basin. GCS would encourage social resilience among citizens instead of approaching the issue from the standpoint of affected communities as passive victims or worse, as well as proceeding with the project of sustainably developing the Mekong Basin without first considering the existing realities of the people in the riparian states. In a nutshell, encouraging social resilience, enabling the self-organisation of people, promoting the capacity of local communities and allowing green values to enter the national discourse are all crucial for the development of adaptive policies to cater to the complex global issues that mankind now faces. Such a perspective would lead to the formulation of a common definition of sustainable development, and aid in a more coherent and cohesive approach to realising the various goals of the UNFCCC.

References

Anderson, Benedict, 1983: *Imagined Communities: Reflections on the Origin and Spread of Nationalism* (London: Verso).

Adger, W. Neil, 2000: "Institutional Adaptation to Environmental Risk under the Transition in Vietnam", in: *Annals of the Association of American Geographers*, 90,4: 738–758.

Adger, W. Neil; Huq, Saleemul; Brown, Katrina; Conway, Declan; Hulme, Mike, 2003: "Adaptation to Climate Change in the Developing World", in: *Progress in Development Studies*, 3,3: 179–195.

Anheier, Helmut; Glasius, Marlies; Kaldor, Mary, 2001: "Introducing Global Civil Society", in: Anheier, Helmut; Glasius, Marlies; Kaldor, Mary (Eds.): *Global Civil Society 2001* (Oxford: Oxford University Press): 3–22.

Baker, Christopher G., 2012: "Dams, Power and Security in the Mekong: A Non-Traditional Security Assessment of Hydro-Development in the Mekong River Basin", NTS-Asia Research Paper No. 8 (Singapore: RSIS Centre for Non-Traditional Security (NTS) Studies for NTS-Asia).

Baker, Susan, 2006: *Sustainable Development* (Oxon: Routledge).

Berke, Philip R.; Conroy, Maria Manta, 2000: "Are We Planning for Sustainable Development? An Evaluation of 30 Comprehensive Plans", in: *Journal of the American Planning Association*, 66,1: 21–33.

Bradshaw, Ted K.; Winn, Karri, 2000: "Gleaners, Do-Gooders, and Balers: Options for Linking Sustainability and Economic Development", in: *Community Development: Journal of the Community Development Society*, 31,1: 112–129.

Brown, Peter G., 2004: "Are There Any Natural Resources", in: *Politics and the Life Sciences*, 23,2: 12–21.

Caballero-Anthony, Mely; Chang, Youngho; Putra, Nur Azha (Eds.), 2012: *Energy and Non-traditional Security (NTS) in Asia* (Heidelberg: Springer).

Chandhoke, Neera, 2001: "The Limits of Global Civil Society", in: Anheier, Helmut; Glasius, Marlies; Kaldor, Mary (Eds): *Global Civil Society 2001* (Oxford: Oxford University Press): 35–53.

Clark, Ann Marie; Friedman, Elisabeth J.; Hochstetler, Kathryn, 1998: "The Sovereign Limits of Global Civil Society: A Comparison of NGO Participation in UN World Conferences on the Environment, Human Rights, and Women", in: *World Politics*, 51,1 (October): 1–35.

Conca, Ken; Lipschutz, Ronnie D., 1993: "A Tale of Two Forests", in: Lipschutz, Ronnie D.; Conca, Ken (Eds.): *The State and Social Power in Global Environmental Politics* (New York: Columbia University Press): 1–18.

Descartes, Rene, 1927: *Automatism of Brutes in Selections* (New York: Charles Scribner's Sons).

Diamond, Larry; Plattner, Marc F.; Chu, Yun-han; Tien, Hung-mao (Eds.), 1997: *Consolidating the Third Wave Democracies: Themes and Perspectives* (Baltimore: Johns Hopkins University Press).

Dobson, Andrew, 2007: *Green Political Thought*, 4th ed. (Oxon: Routledge).

Elliott, Lorraine, 2001: "Environmental Politics", in: Hanson, Marianne; Tow, William T. (Eds.): *International Relations in the New Century: An Australian Perspective* (Victoria: Oxford University Press): 138–157.

Falk, Richard A.; Johansen, Robert C.; Kim, Samuel S., 1993: "Global Constitutionalism and World Order", in: Falk, Richard A.; Johansen, Robert C.; Kim, Samuel S. (Eds.): *The Constitutional Foundations of World Peace* (Albany: SUNY Press): 3–12.

Giddens, Anthony, 1984: *The Constitution of Society: Outline of the Theory of Structuration* (Cambridge: Polity Press).

Greene, Owen, 2005: "Environmental Issues", in: Bayliss, John; Smith, Steve (Eds.): *The Globalization of World Politics: An Introduction to International Relations*, 3rd ed. (Oxford: Oxford University Press): 452–478.

Gregor, Mary, 1996: *Kant: The Metaphysics of Moral* (Cambridge: Cambridge University Press).

Habermas, Jurgen, 1989: *The Theory of Communicative Action, Volume Two: The Critique of Functionalist Reason* (Cambridge: Polity Press).

Hardin, Garrett, 1968: "The Tragedy of the Commons", in: *Science*, 162,3859: 1243–1248.

IPCC (Intergovernmental Panel on Climate Change), 2001: *Impacts Adaptation and Vulnerability: Contribution of Working Group II of the Intergovernmental Panel on Climate Change to the Third Assessment Report of the IPCC* (New York: Cambridge University Press).

IRN (International Rivers Network), 2002: "China's Upper Mekong Dams Endanger Millions Downstream", Briefing paper no. 3 (Berkeley: International Rivers Network).

Karl, Thomas R.; Trenberth, Kevin E., 2003: "Modern Global Climate Change", in: *Science*, 302: 1719–1723.

Laslett, Peter (Ed.), 1988: *Locke: Two Treatises of Government* (Cambridge: Cambridge University Press).

Lipschutz, Ronnie D., 1992: "Reconstructing World Politics: The Emergence of Global Civil Society", in: *Millennium: A Journal of International Studies*, 21,3: 389–420.

McGuigan, Claire; Reynolds, Rebecca; Wiedmer, Daniel, 2002: *Poverty and Climate Change: Assessing Impacts in Developing Countries and the Initiatives of the International Community* (London: Overseas Development Institute).

MEWR (Ministry of the Environment and Water Resources); MND (Ministry of National Development), 2009: *A Lively and Liveable Singapore: Strategies for Sustainable Growth* (Singapore: Ministry of the Environment and Water Resources and Ministry of National Development).

Mortimore, Michael J.; Adams, William M., 2001: "Farmer Adaptation, Change and 'Crisis' in the Sahel", in: *Global Environmental Change*, 11,1: 49–57.

MRC (Mekong River Commission), 2010: *State of the Basin Report* (Vientiane: Mekong River Commission).

MRC (Mekong River Commission), 2013: *Mekong Basin Planning: The Story Behind the Basin Development Plan* (Vientiane: Mekong River Commission).

Muller, Benito, 2002: *Equity in Climate Change: The Great Divide* (Oxford: Oxford Institute for Energy Studies).

O'Brien, Karen L.; Leichenko, Robin M., 2000: "Double Exposure: Assessing the Impacts of Climate Change within the Context of Economic Globalization", in: *Global Environmental Change*, 10: 221–232.

Pogge, Thomas, 2000: *World Poverty and Human Rights* (Oxford: Polity Press).

PRIO (Peace Research Institute Oslo), 1989: *Environmental Security: A Report Contributing to the Concept of Comprehensive International Security* (Oslo: PRIO/UNEP Programme on Military Activities and the Human Environment).

Richards, Michael, 2001: *A Review of the Effectiveness of Developing Country Participation in the Climate Change Convention Negotiations*, Working Paper (London: Overseas Development Institute).

Richards, Michael, 2003: *Poverty Reduction, Equity and Climate Change: Global Governance Synergies or Contradictions?* (London: Overseas Development Institute).

Rostow, W.W., 1990: *The Stages of Economic Growth: A Non-Communist Manifesto*, 3rd edn (Cambridge: Cambridge University Press).

Sachs, Wolfgang; Santarius, Tilman (Eds.), 2007: *Fair Future: Resource Conflicts, Security and Global Justice*, A Report of the Wuppertal Institute for Climate, Environment and Energy (Nova Scotia: Fernwood Publishing).

Sen, Amartya, 1999: *Development as Freedom* (Oxford: Oxford University Press).

Skoufias, Emmanuel, 2003: "Economic Crises and Natural Disasters: Coping Strategies and Policy Implications", in: *World Development*, 31,7: 1087–1102.

Soroos, Marvin S., 1997: *The Endangered Atmosphere* (Columbia: University of South Carolina Press).

Strange, Tracey; Bayley, Anne, 2008: *Sustainable Development: Linking Economy, Society, Environment* (Paris: OECD).

Suryadi, Beni, 2012: "Hydro Energy in Southeast Asia: Issues, Challenges, and Opportunities", in: *ESI Bulletin*, 5,3: 2–5.

Swanson, Darren; Barg, Stephen; Tyler, Stephen; Venema, Henry David; Tomar, Sanjay; Bhadwal, Suruchi; Nair, Sreeja; Roy, Dimple; Drexhage, John, 2009: "Seven Guidelines for Policy-Making in an Uncertain World", in: Swanson, Darren; Bhadwal, Suruchi (Eds.): *Creating Adaptive Policies: A Guide for Policy-Making in an Uncertain World* (New Delhi: Sage Publications): 12–24.

Thomas, David S.G.; Twyman, Chasca, 2005: "Equity and Justice in Climate Change Adaptation amongst Natural-Resource-Dependent Societies", in: *Global Environmental Change*, 15: 115–124.

Tompkins, Emma L.; Adger, W. Neil, 2004: "Does Adaptive Management of Natural Resources Enhance Resilience to Climate Change?", in: *Ecology and Society*, 9,2: 10.

Venema, Henry David; Drexhage, John, 2009: "The Need for Adaptive Policies", in: Swanson, Darren; Bhadwal, Suruchi (Eds.): *Creating Adaptive Policies: A Guide for Policy-Making in an Uncertain World* (New Delhi: Sage Publications): 1–11.

Walby, Sylvia, 2003: "The Myth of the Nation-state: Theorizing Society and Polities in a Global Era", in: *Sociology*, 37,3: 529–546.

WCED (World Commission on Environment and Development), 1987: *Our Common Future* (Oxford: Oxford University Press).

World Bank, 2000: *World Development Report 2000/2001* (Washington, D.C.: World Bank).

Abbreviations

ASEAN	Association of Southeast Asian Nations
BoP	Balance of Payments
CSO	Civil society Organisation

DES	Department of Emergency Services
ESI	Energy Studies Institute
GCS	Global Civil Society
HAPUA	Heads of ASEAN Power Utilities/Authorities
IPCC	Intergovernmental Panel on Climate Change
IRN	International Rivers Network
ISC	Institute for Sustainable Communities
MEWR	Ministry of the Environment and Water Resources
MND	Ministry of National Development
MRC	Mekong River Commission
NGO	Non-governmental organisation
NTS	Non-traditional security
NUS	National University of Singapore
OECD	Organisation for Economic Co-operation and Development
PRIO	Peace Research Institute Oslo
RSIS	S. Rajaratnam School of International Studies
UNFCCC	United Nations Framework Convention on Climate Change
WCED	World Commission on Environment and Development

National University of Singapore

A leading global university centred in Asia, the *National University of Singapore* (NUS) is Singapore's flagship university which offers a global approach to education and research with a focus on asian perspectives and expertise.

Its 16 faculties and schools across three campus locations in Singapore—Kent Ridge, Bukit Timah and Outram—provides a broad-based curriculum underscored by multi-disciplinary courses and cross-faculty enrichment. NUS' transformative education includes programmes such as student exchange, entrepreneurial internships at NUS overseas colleges, and double degree and joint degree programmes with some of the world's top universities, offering students opportunities and challenges to realise their potential. The learning experience is complemented by a vibrant residential life with avenues for artistic, cultural and sporting pursuits. Over 37,000 students from 100 countries further enrich the community with their diverse social and cultural perspectives.

NUS has three Research Centres of Excellence (RCE) and 23 university-level research institutes and centres. It is also a partner for Singapore's fifth RCE. The university shares a close affiliation with 16 national-level research institutes and centres. Research activities are strategic and robust, and NUS is well-known for its research strengths in engineering, life sciences and biomedicine, social sciences and natural sciences. Major research thrusts have been made recently in several fields such as quantum technology; cancer and translational medicine; interactive and digital media; and the environment and water. The university also strives to create a supportive and innovative environment to promote creative enterprise within its community.

NUS is actively involved in international academic and research networks such as the Association of Pacific Rim Universities (APRU) and International Alliance of Research Universities (IARU).

N. A. Putra and E. Han (eds.), *Governments' Responses to Climate Change:*
Selected Examples From Asia Pacific 10, SpringerBriefs in Environment, Security,
Development and Peace, DOI: 10.1007/978-981-4451-12-3, © The Author(s) 2014

Energy Studies Institute

Since its establishment in November 2007, the Energy Studies Institute (ESI) is well on its way towards becoming a presence in the international energy policy research arena. For instance, in the 2012 global go to think tanks index report by the university of Pennsylvania think tank and civil societies program, ESI was listed as among the Top 20 Energy and Resource Policy Think Tanks.

ESI provides timely and quality analyses in an evolving energy research landscape influenced by regional and global events. As a leading energy and resource policy think tank, ESI is committed to leveraging the best resources, methods and tools to carry out policy-related energy research and to raise public awareness. ESI's research have been published in key internationally refereed journals and other publications which includes the ESI Bulletin that is widely read. The researchers at ESI often write newspaper commentaries and appear on television interviews.

As of 2013, ESI has five research tracks, climate change, energy efficiency, natural gas, nuclear energy, and power generation, and two regional networks, Asia-Europe energy policy research network and the East-Asia programme.

ESI Activities

As a thought leader, ESI frequently organises activities that include international conferences and seminars in the areas of energy economics, energy and the environment, and energy security. Its events are organised jointly with other think tanks, government agencies and industries. ESI's activities include the following:

- Perform academic research and undertake energy studies commissioned by public, private agencies and industry.
- Conduct public outreach programmes.
- Participate in the Singapore international energy week and local and overseas energy-related events.
- Conduct research seminars, organise international conferences, workshops, and capacity-building programmes.

N. A. Putra and E. Han (eds.), *Governments' Responses to Climate Change:*
Selected Examples From Asia Pacific 10, SpringerBriefs in Environment, Security, Development and Peace, DOI: 10.1007/978-981-4451-12-3, © The Author(s) 2014

Collaborators and Research Sponsors

Since 2007, ESI has worked with various state and non-state actors such as the following:

- (Singapore government and its agencies) Ministry of trade and industry, ministry of the environment and water resources, ministry of foreign affairs, national climate change secretariat, national environment agency, building and construction authority, energy markey authority, and the land transport authority.
- United Nations economic and social commission for Asia and the Pacific, Asia-Pacific economic cooperation, economic research institute for ASEAN and East Asia, and the centre for non-traditional security studies, S. Rajaratnam school of international studies, Nanyang technological university.

About the Editors

Nur Azha Putra (Singapore) is a research associate with the Energy Security Division at the Energy Studies Institute (ESI), National University of Singapore (NUS), Singapore. He is also non-executive director, board of directors, at the Centre for Research on Islamic and Malay Affairs (RIMA), Singapore. Prior to ESI, Azha was associate research fellow at the Centre for Non-Traditional Security (NTS) Studies, S. Rajaratnam School of International Studies (RSIS), Nanyang Technological University (NTU), Singapore, and a journalist with the national newspaper, Berita Harian (BH), where he received the 'Special Award for Excellence'. Azha graduated from NTU in 2008 with an MSc in international political economy. He holds a Bachelor of information technology from Central Queensland University, Australia (2002), and a professional diploma in risk planning from NUS-Asia institute of risk management (2012).

His latest publications include: Caballero-Anthony, Mely; Chang, Youngho; Putra, Nur Azha (Eds): *Energy and Non-Traditional Security (NTS) in Asia* (Berlin/Heidelberg: Springer, 2012); and Caballero-Anthony, Mely; Chang, Youngho; Putra, Nur Azha (Eds): *Rethinking Energy Security: A Non-Traditional View of Human Security* (Berlin/Heidelberg: Springer, 2012).

Address: Energy Studies Institute (ESI), National University of Singapore (NUS), 29 Heng Mui Keng Terrace, Block A, #10-01, Singapore 119620.
Email: azha@nus.edu.sg
Website: http://www.esi.nus.edu.sg/about-us/our-researchers/nur-azha-putra

N. A. Putra and E. Han (eds.), *Governments' Responses to Climate Change:*   123
Selected Examples From Asia Pacific 10, SpringerBriefs in Environment, Security,
Development and Peace, DOI: 10.1007/978-981-4451-12-3, © The Author(s) 2014

Eulalia Han (Singapore) is a research fellow with the energy security division at the Energy Studies Institute (ESI), National University of Singapore (NUS), Singapore. Prior to ESI, Eulalia was a teaching assistant and tutor at various Australian universities. She was also policy officer at the Department of Emergency Services (DES), Queensland, Australia. Eulalia has a PhD in International Relations (2011) from Griffith University and a bachelor of arts in political science (2007), which she graduated with Class I Honours, from the University of Queensland. She is a recipient of the 'Australian Post-Graduate Award' scholarship (2009–2011). Her latest publications include: (with Rane, Halim): *Making Australian foreign policy on Israel-Palestine: Media coverage, public opinion and interest groups* (Carlton: Melbourne University Publishing Limited, 2013).

Address: Energy Studies Institute (ESI), National University of Singapore (NUS), 29 Heng Mui Keng Terrace, Block A, #10-01, Singapore 119620.
Email: esihne@nus.edu.sg
Website: http://www.esi.nus.edu.sg/about-us/our-researchers/dr-eulalia-han

About the Contributors

Gang He (China) is a PhD candidate with the Energy and Resources Group (ERG) at the University of California at Berkeley. He has worked with the Program on Energy and Sustainable Development (PESD), Stanford university, and the World Resources Institute (WRI). He has a Master's degree in climate and society from Columbia university and a BS in Geography from Peking university. His publications include: He, Gang; Morse, Richard K.; "Making carbon offsets work in developing world: lessons from Chinese wind controversy", PESD working paper #90, March 2010 (Stanford: Program on Energy and Sustainable Development, 2010); Morse, Richard K.; Rai, Varun; He, Gang; "Real drivers of carbon capture and storage in China and implications to climate change policy", PESD working paper #88, August 2009 (Stanford: Program on Energy and Sustainable Development, 2009); and He, Gang; "Chinese Society Confronted with Climate Change", *China Perspectives*, No. 1, pp. 77–82 (2007).

Address: Energy and Resources Group (ERG), University of California, Berkeley,
        310 Barrows Hall, Berkeley, CA 94720, United States.
Email: ganghe@berkeley.edu
Website: http://www.ganghe.net

Harbans L. Bajaj (India): BSc Engineering (Electrical) and MSc Engineering (Power) from Punjab University, Chandigarh, India. He was formerly chairperson, Central Electricity Authority (CEA), and ex-officio secretary, Government of India. He was also technical member, Appellate Tribunal for Electricity (APTEL), and director on the boards of the National Thermal Power Corporation (NTPC), Nuclear Power Corporation of India Limited (NPCIL) and PTC India Ltd. (PTC; formerly Power Trading Corporation of India Ltd.). He is currently a practising management consultant, and is member of the advisory boards of Indian Energy Exchange (IEX), NTPC R&D, and independent director on the board of PTC. Apart from the above, he is also member, governing board, Indian National Academy of Engineering (INAE); life fellow, Institute of Electrical and Electronics Engineers (IEEE); and, fellow, Institution of Engineering and

Technology (IET), Institution of Engineers (India) [IEI], and All India Management Association (AIMA). He is pursuing doctoral studies at the University of Technology, Sydney, Australia. His areas of interest include Indian power sector reforms, sustainability and national energy policies. His publications are: Bajaj, H.L., "Strategies for Indian Power Sector Development", in: *India My Dream*, Interactive Lectures Series (Hyderabad: All India Management Association/Hyderabad Management Association, 2003); Bajaj, Harbans L.; Sharma, Deepak; "Power Sector Reforms in India", Paper presented at international conference on Power Electronics, Drives and Energy Systems (PEDES) [Institute of Electrical and Electronics Engineers, 2006]; Bajaj, Harbans L.; Sharma, Deepak; "Power Sector Reforms in India", Paper presented at the 20th World Energy Congress, Rome, November (World Energy Congress, 2007).

Address: S-451, Greater Kailash Part-II, New Delhi 110048, India.
Email: BAJAJHL@gmail.com

Fitrian Ardiansyah (Indonesia/Australia) is a climate and sustainability specialist. He has over 15 years of experience in the field of environmental economics, natural resource management, integrated spatial and land use planning, sustainable commodities as well as climate change and energy. He is currently finalising his doctoral research at the Crawford School of Public Policy, the Australian National University (ANU), Canberra, Australia. Where climate and energy is concerned, he is a former programme director for climate and energy, WWF Indonesia, and was an expert for the Indonesia Forest Climate Alliance (IFCA) and member of the Indonesian official delegates to the United Nations Framework Convention on Climate Change (UNFCCC). He was previously adjunct lecturer at the Post Graduate School of Diplomacy, Universitas Paramadina, and a member of the executive board of the Roundtable on Sustainable Palm Oil (RSPO). He has received the Australian leadership award and Allison Sudradjat award from the government of Australia. His latest publications include: "Risk and Resilience in Cross-border Areas", in: Elliott, L.; Caballero-Anthony, M. (Eds): *Human Security and Climate Change in Southeast Asia: Managing Risk and Resilience* (New York: Routledge, 2012). He also regularly writes for reputable publishers in Southeast Asia.

Address: WEH Stanner Room #1.38, Crawford School of Public Policy, The
         Australian National University (ANU), 132 Lennox Cross, Canberra
         ACT 0200, Australia.
Email: fitrian.ardiansyah@anu.edu.au
Website: http://fitrianardiansyah.com

Neil Gunningham (Australia) has degrees in law and criminology from the Sheffield University, UK. He is a barrister and solicitor (ACT) and holds a PhD from the Australian National University (ANU). He joined the Regulatory Institutions Network (RegNet) at the ANU in January 2002 and is currently co-director of the national research centre for Occupational Health and Safety (OHS) regulation. He was previously foundation director of the Australian centre for environmental law at ANU, visiting and senior Fulbright scholar at the Center for the Study of Law and Society, University of California, Berkeley, and visiting fellow at the Centre for the Analysis of Risk and Regulation (CARR), London School of Economics. His work on OHS regulation has focused on the mining industry and on the relationship between management systems-based approaches, trust and workplace culture. The insights generated apply to other industry sectors and resonate for other areas of regulation. One particular concern (as of the Asia-Pacific Economic Cooperation [APEC] Ministers Responsible for Mining) has been OHS in the broader Asia-Pacific region. Other research agendas concern climate change governance—examining how individual states and key actors within them, international institutions, and key non-state actors perceive these challenges as well as their negotiating possibilities and options. Publications include: *Mine Safety: Law Regulation Policy* (Sydney: Federation Press, 2007); (with Kagan, R.A.; Thornton, D.): *Shades of Green: Business, Regulation, and Environment* (Stanford: Stanford University Press, 2003); (with Sinclair, D.): *Leaders and Laggards: Next-Generation Environmental Regulation* (Sheffield: Greenleaf Publishing Limited, 2002); and (with Grabosky, P.): *Smart Regulation: Designing Environmental Regulation* (Oxford: Oxford University Press, 1998).

Address: National Research Centre for Occupational Health and Safety (OHS)
Regulation, RegNet, Research School of Pacific and Asian Studies, The
Australian National University (ANU), Canberra ACT 0200, Australia.
Email: neil.gunningham@anu.edu.au
Website: http://www.anu.edu.au/fellows/ngunningham/

Peter Drahos (Australia) holds degrees in law, politics and philosophy and is admitted as a barrister and solicitor. He is a professor at RegNet and holds a chair in Intellectual Property at Queen Mary, University of London. He is a member of the Academy of Social Sciences in Australia. Prior to joining the Australian National University, he was an officer of the Australian Commonwealth Attorney-General's department, where he drafted commonwealth legislation. He has published widely in law and social science journals on a variety of topics, including contract, legal philosophy, telecommunications, intellectual property, trade negotiations and international business regulation. He has served as a consultant to the government, international organisations and international non-governmental organisations. Publications include: *A Philosophy of Intellectual Property*, Applied Legal Philosophy Series (Aldershot-Brookfield, US: Dartmouth Publishing Group, 1996); (with Braithwaite, J.): *Global Business Regulation*

(Cambridge: Cambridge University Press, 2000); (with Braithwaite, J.): *Information Feudalism: Who Controls the Knowledge Economy?* (London: Earthscan, 2002); (co-edited with Mayne, R.): *Global Intellectual Property Rights: Knowledge, Access and Development* (Hampshire: Palgrave Macmillan, 2002); and *The Global Governance of Knowledge: Patent Offices and Their Clients* (Cambridge: Cambridge University Press, 2010).

Address: Centre for Governance of Knowledge and Development, RegNet, Research School of Pacific and Asian Studies, The Australian National University, Canberra ACT 0200, Australia.
Email: peter.drahos@anu.edu.au
Website: http://www.anu.edu.au/fellows/pdrahos/

Nur Azha Putra (Singapore) is a research associate with the energy security division at the Energy Studies Institute (ESI), National University of Singapore (NUS), Singapore. He is also non-executive director, board of directors, at the centre for Research on Islamic and Malay Affairs (RIMA), Singapore. More on the coeditors.

Address: Energy Studies Institute (ESI), National University of Singapore (NUS), 29 Heng Mui Keng Terrace, Block A, #10-01, Singapore 119620.
Email: azha@nus.edu.sg
Website: http://www.esi.nus.edu.sg/about-us/our-researchers/nur-azha-putra

Nicholas Koh (Singapore) is senior officer at the energy and environment research directorate, National Research Foundation (NRF), Singapore. Prior to NRF, Nicholas was an energy analyst with the energy security division at the Energy Studies Institute (ESI), National University of Singapore (NUS), Singapore, and also process engineer at Glitsch Technology Corporation. He has Bachelor of Engineering (Chemical) and Master of Science (Technology Management) degrees from NUS, Singapore.

Address: National Research Foundation (NRF), 1 Create Way, #12-02, Singapore 138602.
Email: Nicholas_Koh@nrf.gov.sg

Eulalia Han (Singapore) is a research fellow with the energy security division at the Energy Studies Institute (ESI), National University of Singapore (NUS), Singapore. Prior to ESI, Eulalia was a teaching assistant and tutor at various Australian universities. She was also policy officer at the Department of Emergency Services (DES), Queensland, Australia. Eulalia has a PhD in International Relations (2011) from Griffith University and a Bachelor of Arts in

Political Science (2007), which she graduated with Class I Honours, from the University of Queensland. She is a recipient of the 'Australian Post-Graduate Award' scholarship (2009–2011). Her latest publications include: (with Rane, Halim): *Making Australian Foreign Policy on Israel-Palestine: Media Coverage, Public Opinion and Interest Groups* (Carlton: Melbourne University Publishing Limited, 2013).

About this Book

This multidisciplinary volume articulates the current and potential public policy discourse between energy security and climate change in the Asia-Pacific region, and the efforts taken to address global warming. This volume is unique as it analyses two important issues—climate change and energy security—through the lens of geopolitics at the intersection of energy security. It elaborates on the current and potential steps taken by state and non-state actors, as well as the policy innovations and diplomatic efforts (bilateral and multilateral, including regional) that states are pursuing. This Brief stems from the assumption that its audience is aware of the consequences of climate change, and will therefore, only look at the issues identified. It provides a useful read and reference for a wide-range of scholars, policy-makers, researchers and post-graduate students.This volume comprises papers that were presented at an international conference on Policy Responses to Climate Change and Energy Security Post-Cancun: Implications for the Asia-Pacific Region's Energy Security, organized by the Energy Security Institute, National University of Singapore, at the Traders Hotel, Singapore, on 18 March 2011. The selected papers have been revised to reflect and include the latest development since 2011 and the comments of two anonymous reviewers.

N. A. Putra and E. Han (eds.), *Governments' Responses to Climate Change:*   131
Selected Examples From Asia Pacific 10, SpringerBriefs in Environment, Security,
Development and Peace, DOI: 10.1007/978-981-4451-12-3, © The Author(s) 2014